Southern Gardener's Soil Handbook

Southern Gardener's Soil Handbook

Dr. William S. Peavy

Pacesetter Press
A Division of Gulf Publishing Company
Houston, Texas

Southern Gardener's Soil Handbook

Library of Congress Catalog Card Number 78-058245
ISBN 0-88415-817-9

Edited by B.J. Lowe
Designed by Leigh McWhorter

Illustrated by
Terry J. Moore

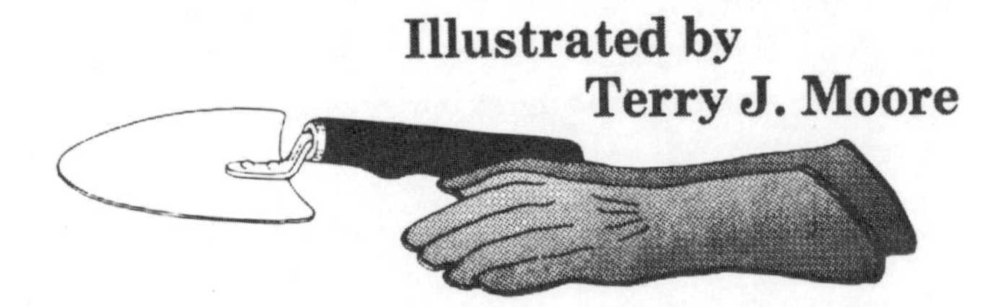

Contents

of special interest...

Acknowledgments

I wish to thank Drs. Charles Welch and Carl Gray of the Soil Testing Laboratory of the Texas A&M University Agricultural Extension Service for helping me unravel some of the mysteries of soil chemistry.

Many thanks also to Bill Adams, Texas A&M Extension Horticulturist, for his encouragement; to B.J. Lowe of Gulf Publishing for his many helpful suggestions; to Leigh McWhorter for her care and skill in designing the book; to Terry Moore for his talented artwork; and to my family for putting up with me the many long hours spent preparing the manuscript and reading proofs.

Hardiness Zone Map

This map shows in moderate detail the expected minimum winter temperatures in the areas comprising the South. Although this book is applicable to the areas within the bold outline, every gardener should check with his or her local agricultural Extension agent for information on variety selection and availability, and special cultural practices necessary for the particular area.

Climate Data for Southern Cities

	Last Spring Freeze	First Fall Freeze	Frost-Free Days	Record January Low (°F)	Minimum Hours of Chilling	Inches of Rain
Alabama						
Birmingham ..	Mar. 19	Nov. 14	241	1	1,000	53
Huntsville	Apr. 1	Nov. 8	221	−9	1,100	50
Mobile	Feb. 17	Dec. 12	298	14	500	67
Montgomery ..	Feb. 27	Dec. 3	279	5	700	51
Arkansas						
Little Rock	Mar. 16	Nov. 15	244	−4	1,000	49
Florida						
Jacksonville ...	Feb. 6	Dec. 16	313	2	400	53
Orlando	Jan. 31	Dec. 17	319	24	300	51
Tampa	Jan. 10	Dec. 26	349	23	200	51
Georgia						
Atlanta	Mar. 20	Nov. 19	244	−3	800	49
Macon	Mar. 14	Nov. 7	240	3	700	44
Savannah	Feb. 21	Dec. 9	291	9	500	48
Kentucky						
Lexington	Apr. 13	Oct. 28	198	−15	1,400	43
Louisville	Apr. 1	Nov. 7	220	−20	1,400	41
Louisiana						
Baton Rouge ..	Feb. 28	Nov. 30	275	10	500	55
New Orleans ..	Feb. 13	Dec. 9	300	14	400	64
Maryland						
Baltimore	Mar. 26	Nov. 19	238	−7	1,400	43
Mississippi						
Jackson	Mar. 10	Nov. 13	248	7	700	50
North Carolina						
Charlotte	Mar. 21	Nov. 15	239	4	900	43
Greensboro	Mar. 24	Nov. 16	237	0	1,100	43
Oklahoma						
Oklahoma City	Mar. 28	Nov. 7	223	0	1,200	32
Tulsa	Mar. 31	Nov. 2	216	−2	1,300	38
South Carolina						
Charleston	Feb. 19	Dec. 10	294	11	600	49
Columbia	Mar. 14	Nov. 21	252	5	700	47
Tennessee						
Knoxville	Mar. 31	Nov. 6	220	−16	1,100	45
Memphis	Mar. 20	Nov. 12	237	−8	1,000	49
Nashville	Mar. 28	Nov. 7	224	−6	1,100	47
Texas						
Austin	Mar. 15	Nov. 20	244	12	700	33
Dallas-Ft. Worth ...	Mar. 18	Nov. 17	244	5	1,000	33
Houston	Feb. 10	Dec. 8	301	19	600	44
San Antonio ...	Feb. 24	Dec. 3	282	0	600	26
Virginia						
Norfolk	Apr. 4	Nov. 9	219	10	1,100	44
Richmond	Apr. 20	Oct. 18	181	−12	1,200	44

Other Quality Pacesetter Books Southerners Will Enjoy

What Soil Is and How to Improve It

Fertile soil is a big factor in successful gardening. And soil fertility depends on three things: physical characteristics (depth, slope, texture, etc.), moisture control, and plant nutrients.

If you understand what an *ideal* soil is like, you are better able to improve your existing soil to make it more like the ideal. Following this, you have to understand how to feed plants and how to irrigate during dry periods.

This book is specially designed to be of real use to *you*, no matter what your background. And it is written to be easily understood by anyone.

Successful gardening is both an art and a science in that it requires practices derived from intuition (a feeling) or from experience as well as practices that are the result of scientific tests or research done with care and precision. Both kinds of gardening practices are effective—and necessary. The main concern is that the practice recommended work and solve the problem, and help us to develop a green thumb without an aching back.

In this book we will use the scientific method as much as possible to get at the facts of gardening, but we'll also rely on the art of horticulture where necessary.

What Is Soil?

Soil is a mixture of tiny mineral particles, air, water, decayed and undecayed organic matter, and living organisms. A *good* soil, such as a silt loam, will (by volume) consist of about half air spaces (pore space) and about half solid particles. About half the air space will be occupied by water (leaving the other half as air space). The 50 percent solids will include a small amount of organic matter.

Northern prairie soils, as in Minnesota, may contain an average of 5 percent organic matter. A sandy southern soil, if crops have been grown in it, is (by volume) likely to consist of about 55 percent solids and 45 percent air space. One percent of the solids may consist of organic matter. This same soil in its virgin state would have lots more pore space and lots more organic matter; it would also be much more fertile, would hold more water and plant food, and water would infiltrate into it much more readily.

Types of Soils

One of the most useful understandings the gardener can have is about the three basic types of soils: sands, clays, and loams. These are referred to over and over again in books, newspaper articles, and at garden meetings in connection with soil improvement and management. These three basic soil types refer to soil texture, or the size of the mineral particles that make up the soil (see Figure 1).

Sands and sandy soils. If you rub two soft stones together for awhile, you will have a small

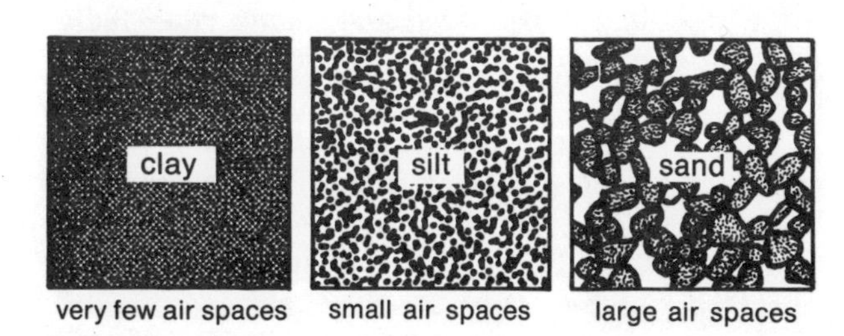

Figure 1. *Relative sizes of soil mineral particles.*

pile of mineral particles or grains. These will be similar to the mineral particles that are predominant in sands, and they are the largest of all soil particles. A sand-type soil will be made up of about 90 percent sand particles.

If you rub some of these particles between your thumb and finger, they will feel gritty. If you wet half a cupful of these particles, squeeze the wet mass into a ball, toss it up a foot into the air and catch it in your hand, it easily breaks apart on the first toss.

These soils are well drained (too much so, in fact) and hold relatively little water or plant foods. They therefore lead to quick wilt or little growth in plants unless they're watered and fertilized frequently. They can be walked on almost immediately after a rain and worked the same day. For this reason some farmers prize them for growing vegetable crops where frequent harvesting is necessary to make the crop pay: mechanized harvesting with tractor-drawn equipment may be done the day after a heavy rain, where in other soil types it might require a week's wait.

Clays or gumbos. These soils are the opposite of sandy soils. They, too, are made up mostly of mineral particles, but the particles are not at all like sand particles. Instead, they are so tiny that an individual particle cannot be seen without a magnifying lens. While a sand particle may be .025 inch (one-fiftieth of an inch) in diameter and can be seen with the naked eye, the clay particle is only .0001 (one ten-thousandth inch), or 250 times smaller than the sand particle (Figure 1).

Clay particles predominate in clay soils, with 45 percent or more of clay and up to 40 percent silt.

Silt refers to the third type of soil particle. You might say it is a medium-size particle, smaller than sand, larger than clay. Silt particles feel "floury" when dry and slightly sticky when wet.

A clay soil, when dry, is very cloddy. When wet, it is extremely plastic and sticky. It is easily molded into various forms—like modeling clay. When pressed into a ball and tossed into the air, a clay ball will not only refuse to break up with the first toss but can be tossed and caught repeatedly without breaking. This is why clay is used to make bricks.

Clays are called "15-minute" soils because after watering you have to wait a long time before you can go into the field to cultivate. And when these soils do get dry enough to work, you have only a short time before they are too hard and cloddy. Farmers may exaggerate this by saying you've got only 15 minutes to get that clay field plowed.

A little clay in any soil is a good thing. In fact, clay can be used to improve a sandy soil so that it holds more water and plant food.

Clay soils "take water" slowly, but once they do they can hold enormous amounts. This is in contrast to a sandy soil, which takes water rapidly but can't hold it.

Clay soils also store large amounts of plant food; a sand can store very little. Wouldn't it be nice if we could "breed" these two together to combine the best traits of each? Fortunately, nature has arranged it so we can do just this. The offspring is called *loam.*

Loams. An ideal soil is a loam with about 45 to 50 percent sand particles, about 35 to 40 percent silt-size particles, and a little clay (10-15 percent). It will weigh about 65 lbs. per cubic foot (when dry). If you take a level cup of this soil, air-dry it and weigh it, it will tip the scales at about 8 oz., or about the weight of an equal amount of water.

But take this same ideal soil and cultivate it for 40 years and "watch it fall apart." A cup will now weigh about 10 oz.! This increase in weight is due to the loss of organic matter and the packing together of the mineral particles due to cultivation.

From the discussion of sand, you remember its excellent ability to absorb water. This is why many potting soils contain liberal amounts of sand.

But a soil that's mostly sand (90 to 100 percent) is not a good soil. Such a soil has too much air space and holds too little water and plant foods. It needs some silt and clay to increase holding capacity for plant foods and water. A cup of sand will weigh about 13 oz.

Nearly as good as a loam is a sandy loam. The difference is that the sandy loam contains more sand (65 percent compared to 45 percent) than the normal loam.

Loams are the most prized agricultural soils in the world—they combine the best characteristics found in sands and clays; they take water readily and store large amounts of water and plant foods for plant growth; they are easily tilled and form few clods.

What Is the Ideal Soil?

The ideal soil is more than just a loam. It must meet several other specifications of structure, depth, slope, and pH.

In general, a superior soil is one that is deep, well drained, gently sloping, well stocked with plant foods, and of medium pH (6.0-7.0).

How deep is "deep"? What is "gently sloping"? See Table 1.

Table 1
Specifications for a Fertile Soil

Characteristics	Description
Texture (fineness of particles)	Loam or sandy loam
Structure (tilth)	Soil in granulated particles.
Organic matter content	5 to 10% (by weight)
Depth	18 in. to permeable lime zone 36 in. to sand or gravel 60 in. to solid rock
Slope (grade)	Uniform slope of 3 to 12 in. per 100 ft.
Depth to water table	60 in. or more
Water infiltration rate	2 to 5 in. per hour*
pH	6.0 to 7.0
Living organisms	Earthworms (2 to 8 per sq. ft.)†
Mineral content: nitrogen (N) phosphorus (P_2O_5) potassium (K_{20}) calcium (Ca) magnesium (Mg)	A rating of "High" in each of the five essential nutrients listed, based on a soil chemical analysis from a standard soil test (see your county agricultural extension agent)
Salinity	Minimal: electrical conductivity less than 2.0 (E.C. x 10^3)
Pore space (air space)	55 to 60% (dry weight of 50 to 68 lbs. per cu. ft., or 6.7 to 9 lbs. per gallon of dry soil)

*This rate is for natural soils. A good potting soil for indoor plants in pots may have an intake rate of 20 inches per hour or more.
†In a natural soil; where synthetic soil mixes are used, earthworms are not necessary.

Most gardeners will probably look at the specs above and say, "But my soil just isn't that good, so why talk about the ideal?" There are things you can do to make any soil more ideal, as long as you know what the ideal is. Maybe your soil can be brought up to those specs.

As a home gardener, you are in a very favorable position to improve your soil, since you're dealing with a limited amount. If you were dealing with an acre (about the size of a football field), it would be a different story. Most home gardeners have perhaps a tenth of an acre (about 4,000 square feet) or maybe only a small area 20 feet by 20 feet (one one-hundredth of an acre).

A tight, sticky soil can be changed to a loose, mellow soil by adding large amounts of organic matter. This a farmer usually cannot do because he has too much land to treat. Later on, we will show you specific things to do to get your soil in shape. Right now, let's look at some of the more obvious things.

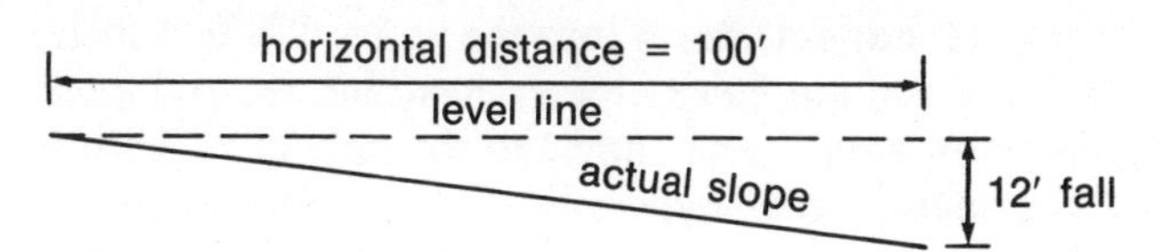

Figure 2. *Maximum tolerable slope per 100 feet in horizontal distance.*

Slope

A gently sloping surface in a gardening area is ideal since it allows surface drainage of excess water but "makes running water walk." (Figure 2). This aids infiltration (soaking in) and prevents erosion. Where the fall is over 12 inches per 100 feet, (1) run rows in a vegetable garden *across* the slope, or (2) build retaining walls and fill in to form level terraces.

Depth

Try to get a soil depth of at least 18 inches. If this is too difficult, go for 12 inches and use drip irrigation (unless you are in high-rainfall country). Be sure to leave "weep holes" downslope in masonry walls. These are small holes to keep water from being dammed up behind the retaining walls; they allow it to drain downward and out of the soil.

Depth to Water Table

The depth to the water table is the distance from the surface down to free water, or the distance you would have to dig down to strike water when digging a well. Shallow wells may tap a water table at 50 feet or less and deep wells may go as deep as 1,000 feet.

The depth to your water table should be 60 inches or more. You can poke a 72-inch soil auger or other narrow stake down and see if you strike soil that is dripping wet. If not, your water table situation is satisfactory.

Water Infiltration Rate

To check this, remove vegetation (grass, etc.) from an area 1 foot square, soak with water until soft, then push a straight sided can (with bottom and top cut out) down into the soft soil 3 to 4 inches. (A tomato juice can is good for this.) Lay a board across the top and tap the can down with a hammer.

On the inside of the can, use a ruler to find the level 2 inches up from the soil surface and mark with felt-tip pen or grease pencil. Fill the can with water up to the 2-inch mark and record the time. Check occasionally. Record the time required for all the water to soak in. If the 2 inches soaked in after 2 hours, your infiltration rate is 1 inch per hour, and this meets our specs in Table 1.

Remember that infiltration rate (intake rate) means the rate at which water moves through wet soil. This is why you have to wet the soil before making the test. A dry soil is much more "thirsty" than a wet one, and it will usually take in water much faster than when it gets wet. A really good soil will continue to take in water readily even after several days of rain; a poor one will not.

If you're checking soil suitability for deeply rooting plants like trees, subsoil drainage can be checked by digging a post hole 2 to 3 feet deep to get into the underlying layers (subsoil). Add water to 1 foot deep. Let drain. Refill to 1 foot and check the time.

Potting soil infiltration rates are a different "ball game." A good potting soil has a fantastic intake rate, and this is necessary for good bottom drainage when you have to deal with a layer of soil only 6 to 12 inches deep. The homemade Peatlite potting soils described on pages 15-16 has an unbelievable intake rate of 30 inches per hour or more. A test of the "Rediearth" brand of Peatlite showed an intake rate of 20 inches per hour; this is because it is ground much more finely than homemade Peatlite. This is typical and shows up in the field where a coarse-textured sandy soil takes water rapidly, a medium-textured soil like loam takes water fairly rapidly, and a clay soil takes in water very slowly.

A test of intake rate of the well-known brand of potting soil "Baccto" showed 12 inches per hour.

A good natural soil will never show the high infiltration rates of a good potting soil, nor is it desirable that it do so. As long as the intake rates are within the range shown in Table 1, the soil should be satisfactory. An intake rate of 4 to 5 inches per hour is ideal for natural soils.

How to Improve Your Soil

If you compare your soil with the ideal soil (Table 1), you may wonder if you really have soil or if you just have the possibility of soil. This is especially true of extreme situations: sand, clay (gumbo), steeply sloping areas, very shallow soils or poorly drained soils. Don't give up! No matter what your situation, you can have a fertile soil. Remember, a green thumb begins with black dirt under your fingernails.

How can soil texture and structure be improved? To find out, have a look at the anatomy of a highly fertile loam soil (Table 2). If you want to make up some soil mixes of your own, the data in this table may be helpful. Using local soil materials it is possible to make a soil mix that will come fairly close to the ideal (Table 3).

Table 2
Physical Characteristics of an Ideal Loam Soil

Material	Composition (%)	Dry Weight (lbs./cu. ft.)	Approximate Volume (gals.)*
Organic matter	(10)	6.5	0.75
Mineral particles	(90)	58.5	6.75
Particle types:			
Sand—50%			
Silt—40%			
Clay—10%			

*7.48 gallons = 1 cubic foot

Table 3
Composition of a Good Soil Mix*
(To make approximately 19 gallons)

Material	Volume (%)	Actual Volume Gals.	Actual Volume Cu. ft.	Dry Weight (lbs.)
Sandy soil	40	7.5	1.0	94.0
Peat moss	60	11.2	1.5	11.2
Total	100	18.7	2.5	105.2

*Based on the desired physical features outlined in Table 2.

Duplicating the Ideal Soil

Well, if you could get the ingredients and a cement mixer, you might be able to do this. The sand could be had (builder's sand), and perhaps the clay. But it is doubtful that silt particles could be found, much less obtained. Raw organic matter (manure, leaves, straw, etc.) decomposes in the soil to a fairly stable form, called *humus*, which is nearly always added in the form of raw organic matter or as compost (humus).

To duplicate natural, fertile soils, you need a supply of organic matter such as dairy or steer manure (not commercial feedlot manure). Even better is a supply of humus (homemade compost, leaf mold). If raw organic matter is to be used it must be ground up into small pieces as is done when old stalks are run through a compost shredder. Manure usually needs no grinding because the animal, when it chews, does the grinding.

Any kind of organic matter is best stored in 30-gallon plastic garbage cans or covered with plastic tarp, or otherwise enclosed. Why? Two reasons: (1) where rainfall is high, if it is compost you are storing, it will suffer losses of plant foods from leaching, and (2) any nice, loose pile of organic matter lying around becomes a maternity ward for the white grub (grubworm) and its mother, the May or June beetle. In screening compost to prepare soil mixes I have seen hundreds of grubs of all sizes in a single gallon of compost that was used by the adult beetle for egg-laying.

You will also need a supply of mineral particles. Good topsoil is best. Topsoil is usually the top layer of soil down to about 12 inches or so. Good topsoil is a mixture of sand, silt, and clay. (Sandy topsoil may be used.) Even builder's sand can be used if nothing else is available. Avoid very coarse or very fine sand.

If you are really ambitious, you can find a clay deposit and dig out a sack or two. Remember not to try to dig out clay when moist, as it will just "gum up" on you. Wait until it is dry and cloddy. You can pulverize the clods with a brick.

Soil Structure (Tilth)

Consider the ideal loam soil described on pages 2 and 3. At its very best, with the proper proportions of sand, silt and clay and with good structure, it weighed only about 65 lbs. per cubic foot. As soils go, this is a light, fluffy soil. But studies show that the same kind of soil cultivated for many years will weigh about 80 lbs. per cubic foot!

What has happened? The soil *texture* (size of the mineral particles) has not changed, but the *structure* has. The soil particles, after heavy cropping, now lie much closer together and there is less pore space (air space). Every gallon of soil now weighs more than it did: it is denser and harder to the touch; it has lost fertility; gone is the fluffy, porous, sponge-like soil mass that was formerly so fertile and yielded bumper crops.

The chief reason for the change in this soil is the loss of organic matter. As organic matter decomposes in a soil it gives off substances which tend to bind the tiny particles of soil into clumps or granules. The presence of large amounts of humus (decomposed organic matter that resists further breakdown) helps maintain this condition. Like a cushion, the particles of organic matter get between the mineral particles, preventing them from sliding so close together.

Arrangement of the soil's particles into clumps or granules is called *granulation.* When this occurs, soil is said to have good structure and be in good tilth. When soil is in poor tilth, the soil surface over newly planted seeds tends to seal up when wetted, and the germinating seeds have trouble pushing through the sealed surface. But if the soil has good structure it does not seal over; it stays friable and seeds have no trouble emerging from the soil.

Chances are good that if the soil has a high organic matter content and an abundance of living organisms, it will also have good structure.

How to Improve the Structure of Your Garden Soil

If your soil is a loam or a sandy loam, you are fortunate. But so often it is a case of sand (too coarse and too loose) or clay (too fine and too tight). What can you do to improve soil structure?

Organic matter. This is the single most important improvement for both tight clay soils and loose sandy soils—it tightens up loose soils and loosens up tight soils.

A 2-inch layer of *sand* may be applied over a clay soil and worked into the top 6 inches to make the clay less tight and sticky. But when watering, water slowly at a rate that the clay can take, since you have treated only the top few inches.

If there is a thin clay layer sitting on deep sand, you've got it made. All that is necessary is to work the top clay layer into the sand below to convert the clay soil to a sandy clay loam. The most practical way to do this is to use a trenching machine. This machine has a rotary wheel with shovels, and as it rotates it eats its way into the soil, making a deep trench. The soil being cut out of the trench can be routed back into the hole and will be a mixture of both the topsoil and that below.

If the first 2-inch layer of sand worked in does not do the job, work in another 2 inches, along with plenty of organic matter, when it is time to prepare a seedbed next year.

Treating a sandy soil with clay. Sands or sandy soils are just the opposite to clays in their problems—they need something to tighten them up. If a nearby clay deposit can be found, haul in a layer of clay and rototill it into a sandy soil.

How much should you add? Little information is available. We know that a loam soil—the ideal—will contain about 45 percent sand, 40 percent silt, and 15 percent clay. But there is no source of silt particles—clay and sand are all we have to work with, so shoot for a mixture of about ⅔ sand and ⅓ clay. Start with a small test plot by spreading on a 2-inch layer of clay and rototilling in. If you mix 2 inches of clay into 4 inches of soil, you will have that mix of about ⅓:⅔.

Liming to improve soil structure. Liming, or adding limestone, is another practice that will improve soil structure, but the soil pH must be below 6.0 because lime is an alkali. Liming supplies calcium, which encourages granulation of the soil particles into clumps.

Worms for tight clay soils. Earthworms are another way to loosen tight clay or gumbo soils. See pages 58-60 for details.

Minimum tillage. This is a good practice to improve soil structure, especially where cultivation (tillage) has been done too often. Tillage stirs the soil, leading to oxidation and loss of organic matter. Tilling the soil when it is a little too wet mechanically forces the soil particles into hard clumps which become clods when dry. This is part of the process used in making bricks, so if you don't want to turn your garden soil into bricks, don't work it too wet!

The less you cultivate your soil the better. Use mulch or a hoe to control weeds. You do not benefit from a dust mulch.

"Slot method" soil improvement. Many vegetable gardeners attempt far too big a garden. They do this to be sure they'll have enough vegetables, but often there's too much production at one time and much goes to waste. It's not how big a garden you have that counts, but how *intensive* and how productive. An intensive-style garden is so productive you don't need a huge garden to do the job.

How big a garden do you need? For the average family, 1,000 square feet should do it. If you will build up your soil and garden intensively, this will take care of the needs of a family of 3 or 4, with plenty of surplus to dry or freeze or can.

A good way to garden intensively is to mark off row centers 40 inches apart and 15 to 30 feet long. Then drive a permanent row stake (1 inch thick by 2 inches wide by 18 inches long) into the ground at the head and foot of each row. If you soak the bottom 10 inches of the stakes in penta for a few minutes and drain and dry thoroughly before driving, these stakes will last several years.

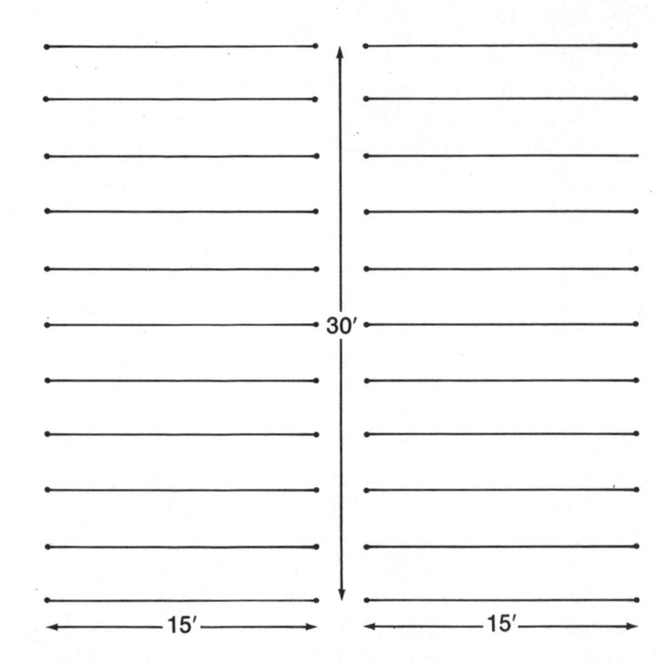

__Figure 3.__ Garden layout of 1,000 square feet for a family of three or four. Each dot is a numbered row stake for using the slot method of soil building.

Don't let the 40-inch spacing scare you. This spacing works best of all. Waste space? No way. For large-size crops like corn and beans, one line of plants per row works fine. For small crops like onions, radishes and carrots, plant two to four lines along the row centers so there is no waste. But, if the 40-inch spacing does scare you, you can get by fairly well with a 36-inch spacing.

Twenty rows, each about 15 feet long will feed a family of 3 or 4 (Figure 3).

To work the slot method of soil building, drive the two row stakes for each row down into the ground until about 7 or 8 inches protrudes up out of the soil. Stretch a piece of heavy cord (like binder's twine) 5 to 6 inches off the ground between the two stakes. Then make a tent-shaped pile of manure under the string just touching the string and the length of the row (Figure 4). Then with a garden rake, flatten the pile of manure to an 18-inch width. This will give you a layer of manure about 2 inches thick. Rototill or spade in the 2-inch layer of manure, but leave the aisle or walkway untouched.

You now have a series of fertile strips 18 to 24 inches wide alternating with untilled walkways. Each fertile strip will accommodate a single line of large-size plants like corn and beans or two lines (twin rows) of small crops like carrots and radishes or up to 4 lines of onions.

Isn't that 2-inch layer of manure too much? No. I have applied up to 3 inches of dairy manure many times with good results. See pages 47-48 for a

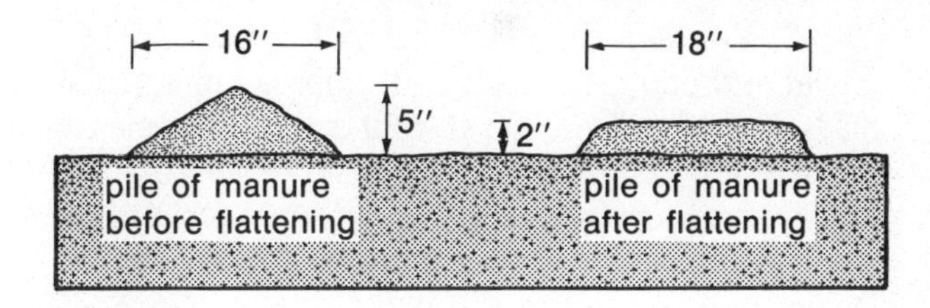

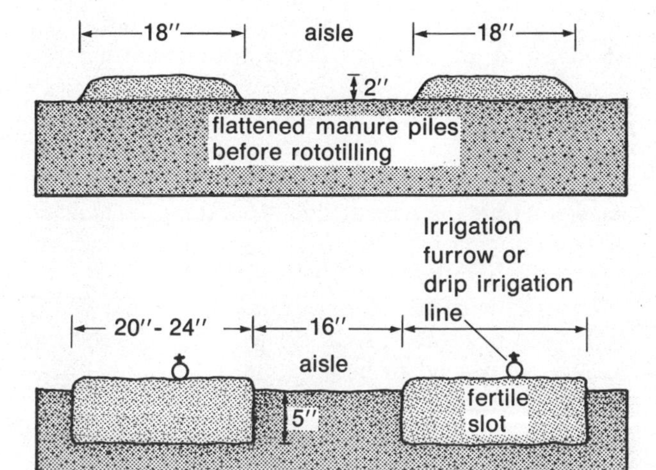

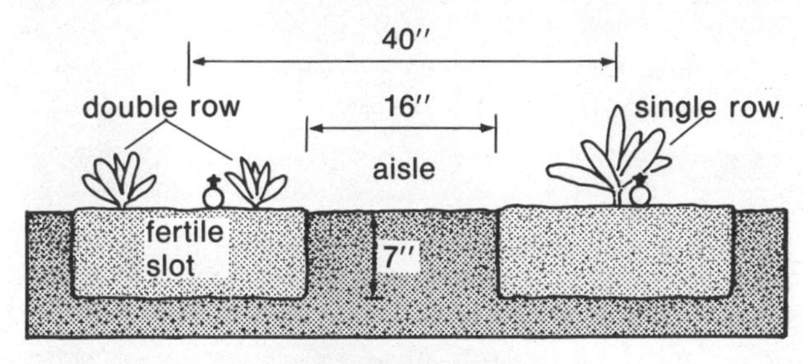

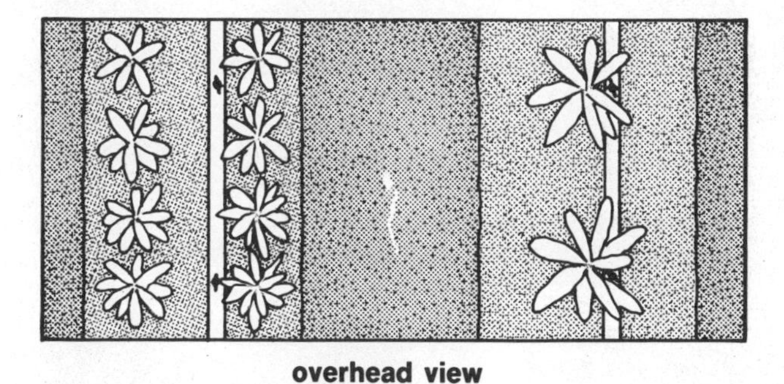

Figure 4. Soil improvement by the slot method.

more thorough discussion. If poultry manure is used, use one-half as much.

How often is the manure treatment repeated? This depends on soil type, average annual temperature, and how much cultivation you do. Do it yearly if you can get the manure; every other year, or every third year for sure. See page 48.

Earthworms. If you keep the soil organic matter content high and use pesticides carefully and sparingly, chances are that earthworms and other beneficial living organisms will move into your garden soil. Mulching with organic mulches will also encourage earthworms.

Spot treatment to improve soil structure. When planting a tree or shrub, "spot" soil treatment may be used to good advantage—that is, you treat a small spot where a plant is to go instead of the whole yard. The idea is to dig a hole a little larger than the root system of the plant, thoroughly recondition the soil that is to go back in this hole, then backfill with the improved soil.

Peat moss, vermiculite, compost, or finely shredded bark may be used as a soil amendment to mix with the soil dug out of the hole (see Mix #2, page 20). Half the volume of the backfill soil should be of the soil amendment, preferably with some organic fertilizer, added too.

For rose bushes, a hole about 12 by 12 inches and 16 inches deep will do; for a larger shrub, a hole 18 inches each dimension; and for a larger tree, up to 2 or 3 feet each dimension.

Roses respond to the spot treatment more than some other shrubs. If the soil is over pH 7.0, a cup of sulfur mixed in with the backfill soil will help acidify it.

To recondition soil for backfilling, do your mixing in a heavy-duty wheelbarrow using a flat-point

When planting a shrub, dig a hole, place the soil on a tarp, and mix the soil, peat moss (and any other amendments) for backfilling into the hole.

A heavy-duty wheelbarrow and flat-point shovel are handy for mixing soil amendments with the soil. To improve structure, vermiculite is the amendment used here.

Using an improved soil mix of half native soil and half organic soil amendment. Pour a little of the mix in the bottom of the planting hole before planting.

Materials for spot improvement of soil when planting shrubs or trees. Organic matter such as peat moss for soil structure and sulfur to adjust the soil pH downward. Peat moss mixed half and half with soil, with a cup of sulfur added, will get the rose bush to be planted here off to a good start in what is otherwise a very alkaline soil.

shovel. You can dig the hole with a round-point shovel and pile the soil on a tarp spread out on the grass near the hole. With your soil amendments nearby, you can throw a shovel full of soil into the wheelbarrow, then a shovel full of amendment and mix it all together like making a giant cake. The sequence can be seen in the photographs at left.

All About Soil pH

Soil pH (soil reaction) is the topic of many discussions at garden clubs and among gardeners. Just what is it and why is it so important?

Soil pH is defined by the soil chemist as "the negative logarithm of the hydrogen-ion activity of the soil." What does all that mean? In a practical sense, pH means how acidic or alkaline the soil is. The scale runs from 0 to 13: 7.0 is neutral; above 7.0 is alkaline; below 7.0 is acidic.

In areas with 35 inches or more of rain per year, soils tend to be acid. In areas with 20 to 25 inches or less, soils tend to be alkaline.

When soils are too acid, essential plant food elements such as nitrogen, phosphorus, and potassium become poorly available to the plant (Figure 5), but the important trace elements such as iron and manganese are more available.

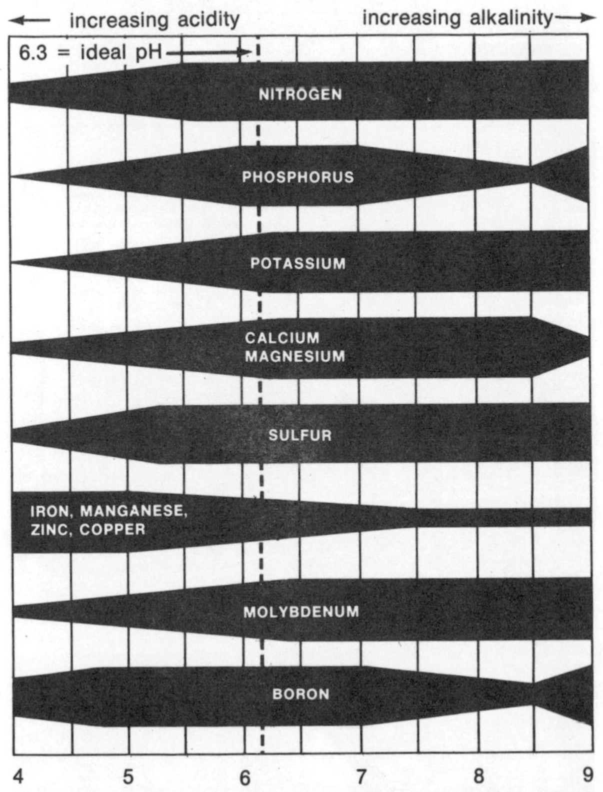

Figure 5. *Availability of the essential elements for plants depends on soil pH. The wider the band, the more available the element. The greatest width of the greatest number of bands falls between a soil pH of 6.0 and 7.0—the ideal is about 6.3.*

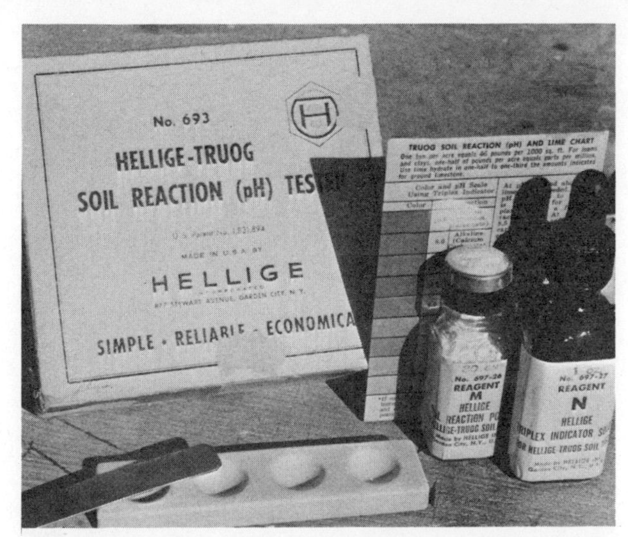

One of the better home garden soil pH test kits. This kit comes with porcelain plates with cavities for soil samples, as well as reagents and a color comparison card.

Making a soil pH test with a home test kit. With the cavity full of soil to be tested, several drops of liquid reagents are added until the sample is wet. The next step is to dust on dry reagent to get a color change.

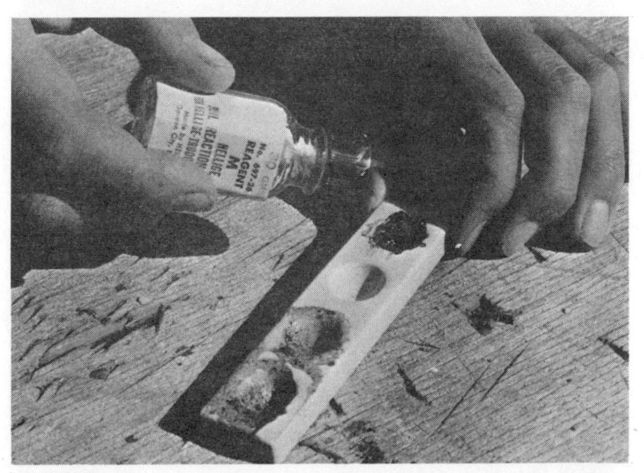

Dry reagent is dusted on the wet sample. After a few seconds there will be a color change, and the color comparison card can be used to match up colors to find the approximate pH.

On the other end of the scale, if the soil pH is too high (around 7.5 or above) other important and essential elements become unavailable or are poorly available, such as phosphorus, iron, zinc, and copper (Figure 5).

If you check the band width of all the essential elements in Figure 5, you will find a point where most of the bands have the greatest width—the point where your soil can do the best job of feeding your plants. This point is a soil pH of about 6.3.

Though the most accurate way to test soil pH is to send it to a lab where a pH meter is available, a home soil test kit does a fairly good job. These are readily available and easy to use (see photos at left). Be sure you dry the soil before testing it.

Raising Soil pH

When soil pH is less than 5.8 or 6.0, it is usually beneficial to add ground limestone to raise the reading to an ideal of about 6.3. You may use regular limestone, but even better is dolomite (dolomitic limestone) because this material not only adds calcium (like the regular limestone) but magnesium as well. The limestone should be broadcast and worked into the soil 6 to 8 inches deep.

The more acid your soil and the finer its texture, the more limestone you'll need (Table 4).

Note that some rates vary widely (as little as 20 lbs. to as much as 40 lbs.). This is because of individual differences in soils. To find if your soil needs the lesser or the greater rate, do a small test plot and check it to see if you raised the pH to the desired point. If not, adjust your rate and then treat your entire plot. A recheck of soil pH about every 3 years should be sufficient.

Lowering Soil pH

Sulfur is one of the most common materials used to lower soil pH. The amount required depends on how much you want to change the pH and how sandy the soil is. The more change you want, the

Table 4
Pounds of Limestone Required per 100 Square Feet to Raise Soil pH of the Top 6 Inches of Soil to 6.5

	Sands (lbs./100 sq. ft.)	Loams (lbs./100 sq. ft.)	Clay Loams (lbs./100 sq. ft.)
Under 5.5	10-15	15-30	20-40
5.5-6.0	5-10	10-15	15-20
6.1-6.5	2-5	5-10	10-15

Table 5
Pounds of 95 Percent Sulfur Required per 100 Square Feet to Lower Soil pH to 6.5

Soil pH	Sands	Loams or Clay Loams
7.5	1.0-1.5	2.0-2.5
8.0	2.5-4.0	4.0-5.0
8.5	4.0-5.0	5.0-7.0
9.0	5.0-7.0	7.0-10.0

Note: 1 heaping pint = 1 pound; 1 heaping gallon = 8 pounds

more sulfur you'll need. But the more sandy your soil is, the less sulfur is required. The amounts of sulfur commonly recommended are shown in Table 5.

Sulfur comes powdered or pelletized. The pellet or flake form is much easier to use and to apply— The powdered form tends to blow away in the wind, cannot be thrown very far from the hopper when broadcasting, and when it gets in your eyes it feels like someone squirted you with acid. The area to be treated should be measured to get the number of square feet and the proper amount of sulfur measured out. The "cyclone" or centrifugal fertilizer spreader used in the photos is a good way to apply the sulfur.

You may find that even the highest rate shown (10 lbs. per 100 square feet) will not lower pH in some soils—silty clay loam in a low rainfall area, for example. In far West Texas a test on such a soil with a beginning pH of 8.5 showed that 30 lbs. of 95 percent sulfur per 100 square feet were necessary to drop the pH to 7.0. Perhaps even more important, no improvement in plant growth was noted and the effect on the soil pH was only temporary. Under these conditions it seems likely that trying to improve the pH of a highly alkaline soil with sulfur is futile.

Before treating your entire garden, first treat some test strips, say, 5 feet long by 2 feet wide (10 square feet). Treat each with 1, 2 and 3 lbs. of sulfur, let incubate in moist, warm soil for 3 months, then test for pH with a home test kit after 3, 6, and 9 months.

Ferrous sulfate may also be used to lower soil pH, but pound for pound it is only about 20 percent as effective as sulfur in developing acidity to lower pH. However, ferrous is a good iron source as well as a source of acidity, so it's worth using if your soil has an iron deficiency.

Where acid-loving plants such as rhododendrons or azaleas are to be grown, acidic organic matter is useful to reduce soil pH. Pine needles, leafmold, tanbark or acidic peat moss may be used.

Applying sulfur (or lime) over a large area can be a headache without proper equipment and materials. To make the job easy, use a two-wheel centrifugal broadcaster and flaked sulfur (not powdered).

Loading a centrifugal broadcaster with flaked sulfur.

Leaching to lower soil pH. High pH usually indicates an accumulation of soluble salts in the soil. Unless a lot of sodium is present in the soil, leaching with good-quality water will help the situation.

Let's take a "salted out" pot plant for example. It has been watered with tap water containing minerals. The water has been used by the plant and evaporated from the soil surface, but few of the salts have been used, so they accumulate. Finally, the plant goes into decline.

The remedy here is leaching, or washing, the accumulated salts out of the rooting zone. To do this, water the plant until a little water comes out the bottom hole. This initial watering will dissolve the salts in the soil, but it doesn't get rid of them, so the plant will continue to suffer even though you watered it with, say, a pint of water. Now add another pint of water and let it run out the bottom hole. The salts that were dissolved before will now be washed out of the soil in the pot.

How do you know if a pot plant needs leaching? Water it until half a cup of water drains out at the bottom. Then test the drainage with a DS meter (Dissolved Solids meter), and if it tests over 3500 ppm (parts per million) dissolved solids, it has begun to salt out and needs leaching (see pages 12-13 for details).

You must have good-quality water for leaching, preferable water low in sodium and with less than 600 ppm dissolved solids.

If your soil contains considerable limestone (calcium carbonate) and you have good-quality water, leaching alone should lower soil pH. To test soil for limestone, pour a little hydrochloric acid (from a local drugstore) or sulfuric acid on a small clod. If it bubbles, limestone is present.

If *no* bubbling occurs, you need to add a calcium-rich amendment like gypsum, and then leach. The gypsum will put injurious sodium salts in a soluble form so they can be leached downward and out of the root zones of plants.

Remember, if you want to lower soil pH and keep it lowered, you not only have to mix in a soil amendment but also follow up by leaching with good water. If your water is alkaline, you won't be able to lower your soil's pH much below that of the water.

Before buying any land, be sure to check the sodium content of the water. If it is high in sodium, *beware.*

Special pH Problems

If your soil test comes back with a pH of 6.0 to 7.0, rejoice—you are "home free": 6.0 to 7.0 is the pH range where most plants grow best.

If you are a perfectionist and your test comes back 6.0, you may want to add just a little limestone to raise the pH to 6.3, probably the ideal soil pH. If your test comes back 6.3 to 6.5, don't touch a thing—your pH is nearly perfect.

But if you live out West, or if for some other reason your test comes back 7.5 to 8.5, you've got problems. And the higher the soil pH is above 7.0, the greater the problem.

The problem is really two-fold: first, there is the amount of soil amendment necessary to lower the pH to 6.5. The pH scale is a logarithmic. This means that a pH of 7.5 is 10 times as alkaline as 6.5 but 8.5 is 100 times as alkaline as 6.5. A pH of 8.5 takes *lots* of amendment, such as sulfur, to correct.

The second problem is that where soil pH is high the water available for irrigation usually has a high pH as well and is highly mineralized. Highly mineralized water, when tested for mineral content, will usually show 1000 to 3000 ppm of salts. Excess salts are what caused the high pH in the first place. Trying to get rid of excess salts with salty water is difficult to do. In the East the 35 inches or so of average annual rainfall leaches out minerals to keep the pH below 7. In the West, the limited rainfall leads to little leaching, to accumulation of salts, and to highly mineralized irrigation water. Every time you irrigate you also add calcium, magnesium, and other base elements. This is equivalent to adding a little limestone every time you irrigate.

In the West, therefore, it is not likely that you can get soil pH below 7, so you just have to put up with it. This means contending with problems of poor phosphorus, iron, zinc, manganese and copper uptake by plants (Figure 5).

This challenge is usually met in two ways, first by growing only those crops that tolerate high soil pH and thrive in spite of it; and second, by foliar feeding (spraying trace element solutions on the leaves).

Soil Salinity

What Is Soil Salinity?

Salinity describes "saltiness." And sometimes a soil is saline (salty) enough to hamper or kill most plants.

Where do salinity problems occur? In outdoor natural soils, salinity is usually a problem where average annual precipitation (rainfall and snow) is 20 inches or less.

But arid areas are not the only places where salinity is a difficulty. If you grow houseplants, no matter where you live you will sooner or later have a salinity problem. Or, you may have a localized soil salinity problem around some plant you have over-fertilized.

Salinity is due to the accumulation of soluble salts in the soil—common table salt (sodium chloride) and its relatives, such as potassium chloride. In high rainfall areas these excess salts are usually washed out of the soil, so there is no problem. But in the arid, low rainfall climate of the West, these salts accumulate.

Occasionally, areas in the humid East have soil salinity problems. If you want to check your soil's salinity, contact your county agricultural agent.

You can have a sample sent to a laboratory equipped with a device that measures electrical conductivity. You can purchase a home meter, too; see "Testing for Houseplant Salinity."

The higher the salt content the higher the current that will flow through the soil sample, and the higher the reading. When the E.C. (electrical conductivity) is 4 or more, you've got salinity problems.

How Did the Salts Get There?

Fertilizers are salts. Read the small print on a bag of inorganic commercial fertilizer. Often you will find something like, "Derived from calcium nitrate, potassium sulfate and ammonium phosphate." Every one of these is close kin to common table salt—sodium chloride—and hence is a salt-type fertilizer.

Salt build-up can even occur when only organic fertilizer is used, though usually not as rapidly. This is because the plant food elements in the organic fertilizer must first be mineralized into salt-type forms before they can be used by plants.

Say, for example, that a fertilizer recommendation calls for 3 cups of salt-type fertilizer per 100 linear feet of row in a vegetable garden, but you double or triple this rate and then plant. Plants may come up and then die. Why? You have "burned" the plants by over-fertilizing. In the medical field this would be like administering an overdose of prescribed medication. If you were a doctor, you might get hit with a malpractice suit.

Perhaps you may have overdone it with a shade tree. I once fertilized a big fruitless mulberry, got interrupted in the middle, forgot where I was, came back and finished the job, but tripled the dose. The leaves began to show salt burn symptoms—leaf margin and tip burn. The leaves began to turn brown and die (see photo). Fortunately, I recognized the symptoms and immediately set a sprinkler under the drip line (the outer limits of the branches) and applied about 6 inches of water to wash the soil. Every one of the leaves fell off, but within a week new growth appeared, and in another 2 weeks the tree was fully leaved. Had I not leached the salts out of the soil the tree would have died.

"Salted-Out" Houseplants

Houseplants may also get salt burn. If the tips or margins are dead and brown, you may have a slight case of salinity in the flower pot. In more severe cases the plant may go into decline and finally die.

Salt burn of pecan foliage.

If you water with rain water, you are less likely to get a salt build-up. But even so, if you fertilize, you are adding salts to your soil. Water evaporates from the surface, and plant leaves draw water out of the soil but leave the salts.

Testing for Houseplant Salinity

The easiest way to see if your houseplants are suffering from salinity is to use an electrical conductivity meter. A good portable meter is the Myron L DS Meter put out by the Myron L. Co., 1133 Second St., Encinitas, CA 92024. The "DS" stands for "dissolved solids." The meter has a small cup at the top that you fill with water and then press a button to see how many parts per million of salts are present. Pure water or distilled water will read zero. Rain water should also read zero or close to it.

To check your plant, first check your leaching water. Say it reads 100 ppm (up to 500 ppm is excellent-quality water provided it does not contain a lot of sodium). Now water your houseplant slowly until half a cup drains out at the bottom. Test it. If it reads 3500 or over, you have salt build-up. Subtract the reading of your water from the soil reading to get an accurate measurement.

Send Your Houseplant to the Laundry

If you find a salt build-up, leach out the excess salt much like you'd launder dirty clothes to get the dirt out. You simply run good-quality water through the root ball to do the job.

If you have access to a DS meter, you can keep running water through (leaching) until the reading of the drainage water drops to a safe point (approximately 1500 ppm). If you don't have a meter, you'll just have to run about 8 cups through and hope for the best.

Preventing Salt Build-Up

Better than letting a plant get sick from salt is to leach a little every time you water. This means always watering until a little water runs out the bottom of the pot. A rule of thumb is to measure the amount of water it takes to get a few drops out at the bottom, then add 20 percent more for leaching. For example, if a pot plant takes a cup of water to irrigate to completion (a few drops out at the bottom), add another ¼ cup for leaching.

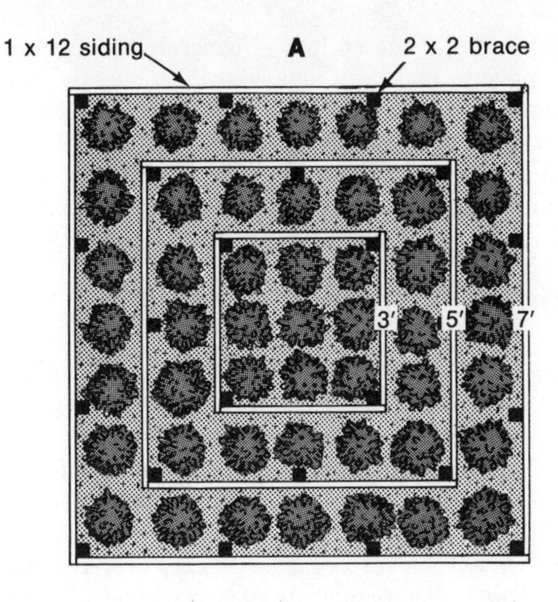

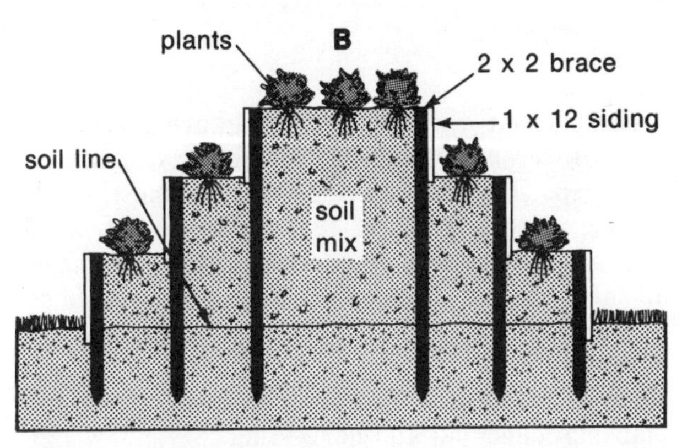

Figure 6. Pyramid structure for intensive gardening. A is a top view; B, a side view.

Container Gardens and Potting Soils

Pyramid Container Gardening

One of the most interesting and efficient methods of container gardening is the pyramid-shaped planter box (Figure 6). Instead of a long, rectangular box you build a square or a round structure with steps that go upward, and on every step there is a planting space.

You'll need plenty of edging material for building the pyramid structure. The most practical way is to build the bottom edging of thin concrete blocks in a square shape; then build succeeding levels using corrugated metal or plastic lawn edging. This edging is ⅛-inch in thickness or less, so it gives you maximum space for growing things. Otherwise, you could go all the way up with concrete blocks.

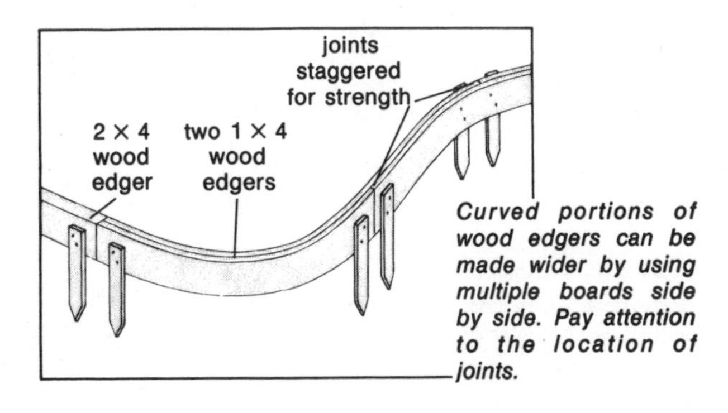

Curved portions of wood edgers can be made wider by using multiple boards side by side. Pay attention to the location of joints.

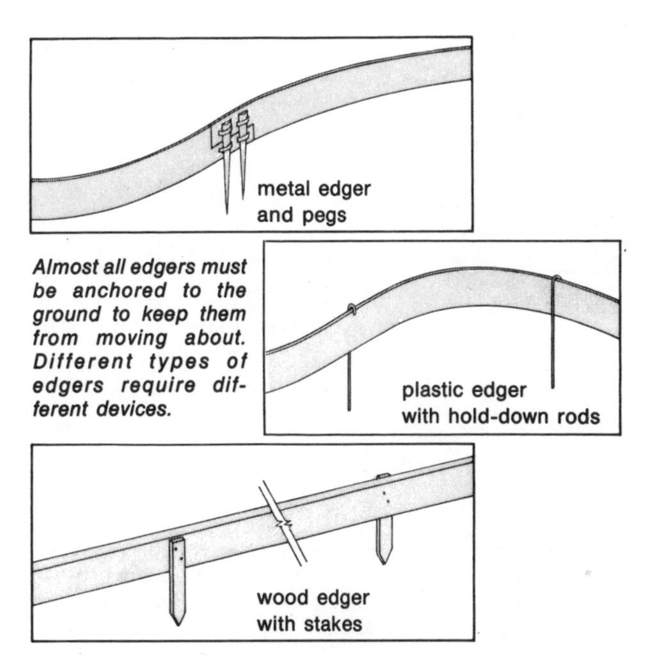

Almost all edgers must be anchored to the ground to keep them from moving about. Different types of edgers require different devices.

If you build the bottom of concrete blocks, you will have to make one side 48 inches long and the other 56 inches (Figure 6) to make the blocks come out even. Fill this with a good soil mix to within 1 inch of the top. Then move in about 6 inches and make a circular ring of curbing, pushing the curbing down about an inch into the level below to secure it.

Planting suggestions for the three levels:

Level A—7 inches deep; small, shallow-rooted crops like lettuce, carrots, onions, radishes, leeks, chives, parsley.

Level B—14 inches deep; medium-size, medium-rooting depth crops like beans, dwarf melons, dwarf tomatoes like Patio and Toy Boy.

Level C—21 inches deep; large-size, deep-rooting crops like regular tomato varieties Spring Giant, Big Set, Terrific, Fantastic or Small Fry; kale.

Notice that the pyramid structure gives you three different rooting depths: shallow, medium, and deep. The plant growing areas are mostly concentric circles of narrow strip of soil about 6 inches wide. This may seem crowded, but remember: all the space is available for plant growing since no walkways are needed. It is wise to limit the maximum width of this structure to 56 inches or less, since you can reach 24 to 30 inches comfortably for planting, checking and harvesting.

In filling one of these pyramids you will find a concrete mixer helpful, since plenty of soil mix is needed. In Figure 6 the first level alone takes about 58 gallons of soil mix; level B takes about 29 gallons; and level C takes about 14 gallons, for a total of about 100 gallons.

To fill this pyramid, use a mixture of half sand or loam mixed with about half soil amendments plus some compost or manure (Mix #2, 3, or 4, page 20).

What can you grow in these pyramid gardens? Almost anything except very large plants like corn, okra, squash and the like. Flowers grow well in them. So do strawberries, and if the strawberry plants are planted in the center of the planting strip 3 inches from the curbing, the fruits will hang over and stay nice and clean since they won't touch the soil surface.

Herbs also do well in this garden. You can use the lower level to grow more than enough herbs for the kitchen.

If filled with rich soil, the pyramid garden will yield bumper crops. And since it is up off the ground, it takes less bending and stooping to cultivate.

The Hollow Block Container Garden

An effective but inexpensive planter box can be made by placing hollow cement block or tiles with open end up on a patio or whatever surface is available (Figure 7). Fill the cavity with potting soil to within an inch of the top (to leave a space for watering) and then plant. For deeper-rooting crops, try stacking blocks 2 to 3 deep.

Figure 7. *Cinderblocks make interesting dividers for outdoor plantings. They can be set above ground to line a walk or patio or to border a raised bed; buried in the ground, they make attractive designs and facilitate culture of plants with different requirements.*

Potting Soil vs. Natural Soil

Natural soil, even if it's good, does not make a good potting soil. The only exception would be a planter box that is (1) to be permanently located and does not have to be moved and (2) 18 inches or more deep.

There are several reasons why even good natural soil does not make a good potting soil. One is that when you are moving or handling pots of soil, you don't want to break your back doing it. Natural soil will weigh 10 to 15 lbs. per gallon when moist, while specially prepared potting soils will weigh from 3 to 8 lbs.

Another reason natural soil is not good for most container gardening is that the soil depth is too shallow for good drainage. The lower part of the soil in a pot will always tend to be saturated long after irrigation because the pore spaces are small and gravity is not strong enough to pull all the excess water down and out. Even the lightweight factory-made peatlite may have this problem. Due to large pore spaces, the home-mixed peatlite does not have this problem—it drains perfectly. (See the following sections.)

Since sand has the largest pore spaces of any natural soil, it is often used as a part of a potting soil. But it should be medium sand; fine sand has very fine particles and tiny pore spaces, so it does not absorb water readily or drain well. However, nearly any kind of sand can be used in a potting soil if mixed with soil amendments, especially vermiculite.

In a test of water absorption and time to drain, using a 6-inch flower pot it took sand 8 minutes to soak up a pint of water and stop draining. When the same sand was mixed half-and-half with sphagnum moss, it took 4 minutes. When mixed half-and-half with vermiculite, it took only 2 minutes. The vermiculite used was horticultural (coarse) grade #2.

Synthetic Soil: The Cornell Peatlite Mix

A gallon of dry sand will weigh about 11 lbs. and hold about 30 percent more than that (3.3 lbs., or 1.6 quarts) in water when irrigated. A mix of half sand and half sphagnum moss will weigh about 7 lbs. per gallon dry and will absorb about 43 percent of that weight in water when you irrigate. But *peatlite* weighs less than 1 lb. per gallon dry yet absorbs 258 percent of its dry weight in water when irrigated!

Peatlite was developed by Dr. Shelldrake of Cornell University and is sometimes called the "Cornell Mix." It is made up of half shredded sphagnum moss and half vermiculite with a small amount of plant food elements added. Peatlite is a contraction of the names of its two parents, *Peat*-moss and vermicu*lite*.

Recipe for Cornell Peatlite

Sphagnum moss (shredded)	88	gals.
Vermiculite (Horticultural Grade #2 or 4)	88	gals.
Dolomitic limestone	5	lbs.
Superphosphate (0-20-0)	1	lb.
5-10-5 commercial fertilizer	12	lbs.

The above formula contains enough plant food for about 6 weeks. After this, feeding must be done unless timed-release fertilizers plus micronutrients are added).

Potting Soil for Houseplants

We have talked about food plants in containers. What about all the houseplants you want to grow? What kind of potting soil is best for them?

Most potting soils leave a lot to be desired for houseplants. With many of them, all the excess water will not drain out after watering. The taller pots do a little better, but the very shallow pots may have a continual condition of "wet feet."

What about commercial potting soil and peatlite? Most commercial potting soil and factory-made peatlite need some help before they give best results with houseplants. The pore spaces in these soils are too small. To make them better, mix half-and-half with vermiculite, but be sure it's the coarse #2 grade.

Home-Mixed Peatlite—A Superior Potting Soil

Home-mixed peatlite (Table 6) is a superior potting soil, better than anything you'll find in the store, especially for houseplants in containers. Don't take my word for it—try it yourself. Buy two of the same kind of plant about the same size. Leave one in the original soil but repot the other in the home-mixed peatlite. Treat both the same and check the growth after 6 months. Irrigate one of your 6-inch pots with a pint of water and notice how quickly it soaks in and how long it takes to stop draining. Now do the same with a 6-inch pot filled with peatlite. You will find the water sinks down into the soil in the peatlite pot almost immediately and is all finished draining in a very short time. This means that all the extra water near the bottom of the peatlite pot has drained out.

Table 6
Formulas for Modified Peatlite Mix

	Ingredients		
Volume	Sphagnum Peat Moss (Shredded)	Vermiculite (Horticultural Grade #2)	Dolomite (Dolomitic Limestone)
Small (½ gal.)	1 qt.	1 qt.	2 teaspoonsful (½ oz.)
Medium (30 gals.)	15 gals. (2 cu. ft.)	15 gals. (2 cu. ft.)	2¼ cups (1.8 lbs.)
Large (1 cu. yd. or 202 gals.)	101 gals. (13.5 cu. ft.)	101 gals. (13.5 cu. ft.)	7½ pints (12 lbs.)

Note: The above mix contains only two of the 13 essential plant food elements. These must be added for successful plant growth (see pp. 23-26).

When you go to your garden center, be sure you get the coarse Horticultural Grade #2 vermiculite; you will know it by its particle size of ⅛ to ¹⁄₁₀ inch. They will be almost as large as kernels of unpopped popcorn.

Be sure you also get shredded sphagnum moss. The package may be labeled "Sphagnum Peat Moss." If you can't buy dolomite locally, you can order a 5-lb. bag through most nurseries.

The peatlite shown in Table 6 is lacking in 11 of the 13 essential plant food elements, so these must be added in some way such as compost, timed-release fertilizer (such as Osmocote 14-14-14, using 1 to 2 tablespoonsful per gallon) plus a little manure, compost, or cottonseed meal for micronutrients; or nutrient solution such as Steiner can be used.

Although peatlite is especially good in small pots for indoor houseplants, home-mixed peatlite is an excellent potting soil for just about any pot or any plant. It's great for growing your own transplants from seed in paper cups on a window sill.

Small batches of home-mixed peatlite are easily made up in batches of as little as half a gallon (Table 6). A shallow bowl or pan and a small shovel are handy in mixing.

After blending the three ingredients, continue to mix while adding water. You have the right amount of water when you *firmly* squeeze a handful of the moist peatlite and a few drops of water are pressed out. If a few drops come out when you squeeze a handful *gently,* it is too wet.

After wetting the home-mixed peatlite you're ready to go to pot; if not to pot, then to a plastic bag or plastic garbage can to store until ready to use.

For larger batches, use a wheelbarrow or cement slab and a flat-point shovel. For huge batches, a cement mixer is easiest.

Early cucumber transplants being grown on a south-facing window sill in a homemade peatlite soil and watered with Steiner solution.

Ingredients for the superior homemade potting soil—peatlite. Use dolomitic limestone, sphagnum peatmoss, and coarse-grade horticultural vermiculite.

Potting Soil Mixes Compared

If you have abundant sand and a desire to save money, you may be tempted to just fill a container with sand and some timed-release fertilizer and start growing plants. I can tell you from a number of trials this does not work well in container gardening, except perhaps in a planter box 18 to 24 inches deep with some manure added (this is similar to Mix #1, page 20). A 3-quart container filled with sand alone takes water slowly, drains slowly and poorly, and gives poor plant growth. Besides, sand is extremely heavy when it's the sole material in a pot.

A gallon pot of sand just after irrigation will weigh about 14 lbs.; a more desirable mix of ½ sand and ½ sphagnum moss will weigh only about 10 lbs.; and an even better mix of ⅓ sand and ⅔ sphagnum weighs only half again as much.

Vermiculite (#2 grade) makes large pore spaces in soils and allows quick and complete drainage. A mix of ½ vermiculite and ½ sand resulted in complete drainage within 2 minutes in a 6-inch pot.

Peatlite is especially well adapted for use where potting soil has to be lightweight, and where is this more true than with hanging baskets? You can lift a 1-gallon peatlite hanging basket down off its hook and lift only about 3 lbs. If you tried the same thing with mostly sand, you'd be lifting about 14 lbs.!

The other situation where peatlite is so ideal is for pot plants which have a rooting depth of 4 to 5 inches (6-inch pot) or 7 to 8 inches (10-inch pot). The shallow depth sets up your pot of soil for poor bottom drainage unless you use a soil with "king-size" pore spaces. Peatlite has these king-size pore spaces that drain so quickly and completely.

Farming in Flowerpots

Trapped in an apartment or on a lot with no soil? Don't give up—there's a way out. Try container gardening on your patio. All you need is a container such as a pot, bucket or can, some potting soil, and a spot that gets several hours of sun each day.

Container gardening can even be done indoors. Every houseplant is a form of container gardening, or farming in a flower-pot. And you can even grow your own groceries in containers, too.

The world's smallest garden is a plastic cup filled with potting soil. Place it on a sunny windowsill and plant with small-size crops like herbs or onion sets or radishes. (Onion sets are small, immature onion bulbs about the size of cherries. If you place your cup of onion sets in a sunny window and keep them watered, you can have fresh, green onions in a month.)

Using an 8-oz. plastic or Styrofoam cup, punch several drain holes on the side of the cup at the bottom. A heated nail easily burns holes in plastic cups; use a sharp pencil for Styrofoam cups. Fill with good potting soil, and with your finger punch four to five holes an inch deep in the soil and drop an onion set in each hole. Push the set down another inch, cover and water.

Radishes also mature in about a month but are started from seed. Using the eraser end of a pencil, make planting holes in the cup of soil ½ inch deep and ½ inch apart. Start eating the radishes when the roots grow to ½ inch in diameter.

Varieties for Container-Grown Vegetables

Tomatoes:	Patio, Pixie, Tiny Tim, Saladette, Stakeless, Atom
Peppers:	Yolo Wonder, Keystone Resistant Giant, Canape, (Hot) Red Cherry, Jalapeno
Eggplant:	Florida Market, Black Beauty, Long Tom
Squash:	Dixie, Gold Neck, Early Prolific Straightneck, (Green) Zucco, Diplomat, Senator
Leaf Lettuce:	Buttercrunch, Salad Bowl, Romaine, Dark Green Boston, Ruby, Bibb
Green Onions:	Beltsville Bunching, Crystal Wax, Evergreen Bunching
Green Beans:	Topcrop, Tendergreen, Contender, (Pole) Blue Lake, Kentucky Wonder
Radishes:	Cherry Belle, Scarlet Globe, (White) Icicle
Parsley:	Evergreen, Moss Curled
Cucumbers:	Burpless, Early-Pik, Crispy

Note: For additional information on variety selection, consult your county Extension agent or Extension horticulturist.

Planting Information for Growing Vegetables in Containers

Crop	Number of Days for Germination	Number of Weeks to Optimum Age for Transplanting	General Size of Container	Amount of Light* Required	Number of Days from Seeding to Harvest
Beans	5-8	—	Medium	Sun	45-65
Cucumbers	5-8	3-4	Large	Sun	50-70
Eggplant	8-12	6-8	Large	Sun	100-130
Lettuce	6-8	3-4	Medium	Partial shade	45-60
Onions	6-8	6-8	Small	Partial shade	80-100
Parsley	10-12	—	Small	Partial shade	70-90
Pepper	10-14	6-8	Large	Sun	90-140
Radish	4-6	—	Small	Partial shade	20-60
Squash	5-7	3-4	Large	Sun	50-70
Tomato	7-10	5-6	Large	Sun	90-130

*All vegetables grow best in full sunlight, but those indicated will also do well in partial shade.

Sources of Containers

Flower pots of all sizes make good homes for plants if pot size and plant size are matched up. They can be highly decorative too. But if you aren't fussy about looks, many waste products can be recycled from the garbage can for use as containers. Likely candidates are empty half-gallon or gallon milk, soap or bleach jugs. Be sure to wash them out *thoroughly* before using. Cut the tops out and fill with moist potting soil to within an inch of the top. Firm the soil slightly while filling to prevent settling later. Punch several drain holes.

Drainage Holes for Containers

Drainage holes must be made in all containers, but instead of puncturing the bottom of the container, make the holes on the *side*, near the bottom, so you can see when you've finished watering. Make one hole for each 4 inches of circumference around the bottom of the pot.

With plastic cups or jugs a good way—often the only way—to make drain holes is by melting a hole with a hot nail. Find a large nail, preferably a spike ⅛-inch in diameter or larger. Grasp the head end of the nail with a pot holder or pliers and heat the other end until hot.

What Grows Best in Containers?

Tomatoes are the most popular food crop for containers. The Toy Boy variety and the Patio varieties are good for 1- to 2-gallon containers. Herbs are popular container food crops too. Dwarf varieties of many vegetables can also be grown in small containers and regular varieties of naturally small vegetables as well.

Tomato plants grown on a patio in containers filled with homemade peatlite and fed with Steiner solution. Dwarf varieties grow in coffee cans; large types grow in bushel baskets.

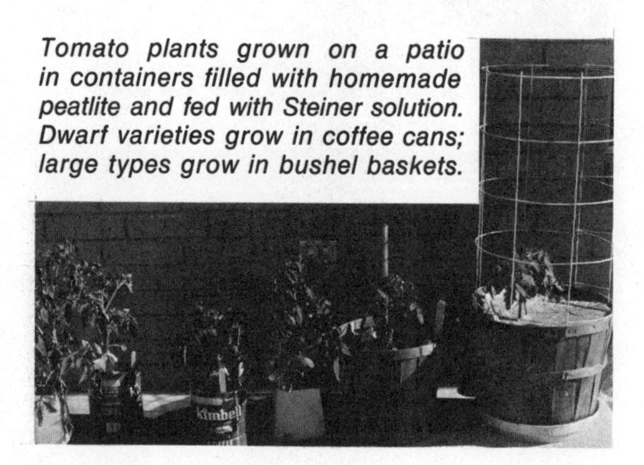

The Mini-Garden

Though you can actually garden in an 8-oz. paper cup on a window sill, you will be very limited in your plant selection. An empty 3-quart container is about the smallest practical size for vegetable gardening in containers. A 3-quart metal coffee can is a good way to get a container free of charge. The small size of this container will limit you to a dwarf-type plant; the Tiny Tim or Pixie tomatoes will fit nicely.

To support the plant, push a thin board (like an old yardstick) down in the center of the pot leaving 2 feet protruding. Tie the tomato stem to this stake to keep the plant from falling over.

For drainage, take a hammer and nail and punch three to four holes on the sides near the bottom.

About the largest container that is practical is a 4- to 5-gallon size. This will accommodate any of the larger tomato varieties, but the Small Fry variety is hard to beat. In my test garden I tried container sizes from 3 quarts up to 8 gallons (photo above). The 8-gallon was larger than necessary and

Heavily loaded "Toy Boy" tomato growing in a half-bushel basket of homemade peatlite.

took too much potting soil to fill. A circular wire cage is handy to support the plant. This does away with the need for a center stake and for pruning, since you must prune to a single stem when supporting a tomato by staking.

Basket Cases

A 4-gallon (half bushel) basket is great for tomatoes and other crops. If you can find a large, shallow drain pan and have space in front of a large south window, you can grow food during winter in your home. The Toy Boy variety of tomato grows well in these.

For cucumbers, use self-pollinating varieties such as Tosca or Gemini. To pollinate indoor tomatoes, shake the plants every day or two when blooming. The blossoms remain open 3 days, so you must shake the vines at least every third day for good fruit set. Outdoor tomatoes get enough shaking from wind.

A plant in a 4-gallon basket can be supported by a piece of surveyor's lath (1½ inches wide x ½ inch thick x 4 feet long) thrust downward in the center of the pot. Builder's supply stores and hardware stores should stock this lath or be able to get it for you.

Line Baskets with Plastic

If you use wood baskets for containers, you must line the inside some way to seal against water loss.

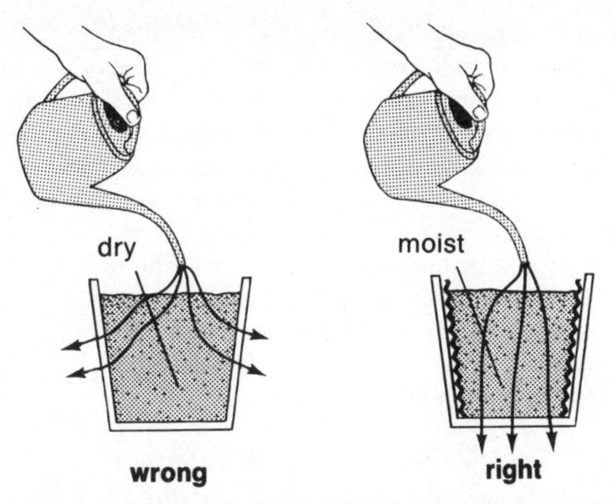

Figure 8. *Lined and unlined baskets used for planters. Plastic-lined basket at right stays moist. Water escapes through sides of the unlined basket (left).*

Polyethylene plastic sheeting is a good material to use. If you do not seal the baskets, when you water the plants, the water will not run into and through the soil as it should. Instead, it will run out through the cracks in the sides of the baskets (Figure 8). By lining the insides with plastic, the water is trapped and must flow downward and through the soil to finally drain out at the bottom.

Cherry tomatoes in hanging baskets are attractive as well as practical.

When using larger baskets such as the 4-gallon size, use one of the lighter-weight potting soils. A 4-gallon basket of sand when moist would weigh about 56 lbs.; a mix of ⅓ sand and ⅔ sphagnum moss would weigh only about 30 lbs. Four gallons of the super lightweight peatlite when moist weighs only about 11 lbs. (but a strong wind could blow it over if a large plant is growing in it). The ½ sand and ½ sphagnum or ⅓ sand and ⅔ sphagnum might be preferable for the needed weight.

Planter Box Container Garden

For small, shallow-rooted crops such as lettuce, carrots, or radishes, a rectangular box about 12 inches wide by 12 inches deep by 4 to 8 feet long makes a good home. Two rows can be planted if each row is planted about 2½ inches in from each side of the box.

To grow larger, deep-rooting crops like tomatoes, beans and peppers, make the box 12 to 18 inches wide and 18 to 24 inches deep.

Potting Soil Mixes

The soil mixes given below are for filling planter boxes, for container gardening, or for gardening in areas behind retaining walls on rocky hillsides where there is little or no soil. The mixes are designed for season-long use without feeding. Amounts are by volume.

Procedure: mix, moisten, set in a warm place, and allow to incubate a month before planting.

Mix #1
2 parts sand or loam (66%)
1 part manure (33%)
¼ cup 0-20-0

Mix #2
3 parts sand or loam (43%)
3 parts S.A.* (43%)
1 part manure or cottonseed meal (14%)

Mix #3
4 parts sand or loam (44%)
4 parts S.A.* (44%)
1 part compost (12%)

Mix #4
4 parts sand or loam (40%)
5 parts S.A.* (30%)
1 part clay (10%)
2 parts manure (20%)

Note: Before adding manure to mixes, add 2 oz. superphosphate (0-20-0) per gallon of manure. If soil pH is under 6.0, add 1 level tablespoonful of limestone per gallon of mix.

*Soil amendment such as vermiculite, peat moss, or perlite. These have no fertilizer value, but they improve soil structure. Manure does both.

How to Make Your Soil Fertile

Simple Soil Chemistry and Plant Physiology for the Non-Scientist

When you go out to feed (fertilize) your plants by broadcasting fertilizer, you're really feeding your soil. The fertilizer first gets into the soil and is then translocated from soil to plant. Only when you spray trace elements such as iron on plant leaves can you transfer food directly to a plant. (This is called *foliar feeding.*)

Plants have tiny mouth-like openings on the undersides of their leaves where carbon dioxide (CO_2) is absorbed and used for growth. But nearly all other elements used in growth are taken in by the roots, and the kind of root that absorbs most plant food is a tiny, threadlike root called a *root hair.* These root hairs absorb liquids only.

The soil is a storehouse for plant food. It is like a kitchen with two pantries: one tiny and limited; the other, huge. The tiny pantry stores limited amounts of plant food and is called the *soil solution.* The other pantry stores large amounts of plant foods by electrical attraction; it is composed of *soil colloids.* Soil scientists refer to the ability of soil colloids to attract and store plant food elements as *exchange capacity.*

Soil Solution

Plant roots cannot feed upon solids. They absorb liquids only, so a solid fertilizer must first become liquid to get into the plant. This is why you must water fertilizer into the soil.

The soil is honeycombed with tiny air pockets called *pore spaces* or *air spaces.* When the fertilizer solution soaks downward into the soil, it goes into these pores until about half the pore space is occupied by the water-and-fertilizer solution. Once inside the pores of the soil, it is called the soil solution.

The soil solution is like millions of drops of water with air bubbles in the center of each drop; each drop is surrounded on the outside by soil particles (Figure 9).

A loam soil will have more pore space than a sandy soil—it will hold more soil solution.

The plant root feeds upon phosphorus in the soil solution in the form of the ions $H_2PO_4^-$ and HPO_4^{-2}. An *ion* is simply an element attached to some other elements, and it carries an electrical charge. Note that the element phosphorus is carried in combination with hydrogen and oxygen as an ion with a

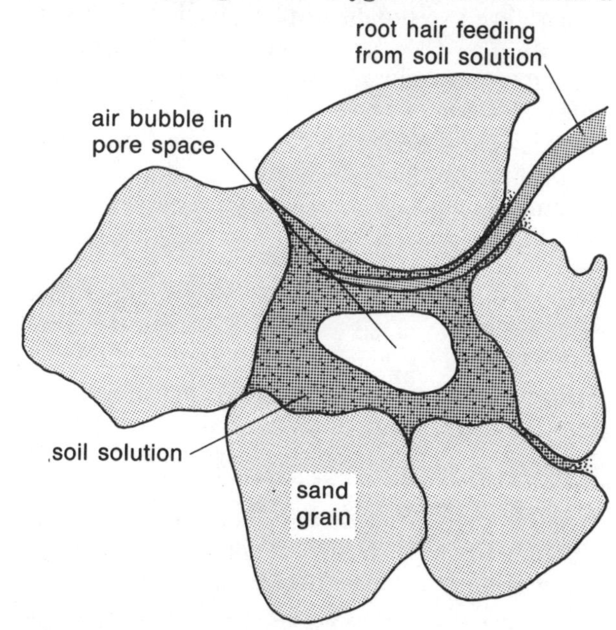

Figure 9. *Plant root feeding from soil solution of a sandy soil (schematic not to scale).*

negative charge. This is in contrast to other elements, such as potassium (described below), that carry positive charges.

The plant root also feeds upon sulfur and nitrogen in the soil solution. These exist as the ions SO_4^{-2} and NO_3^- in the soil solution, where they are very mobile (they get around a lot). They are "here today, gone tomorrow" elements.

Soil Colloids

A big factor in the superior fertility of loam soils is their large plant food storage capacity due not only to lots of pore space but also to tiny solid particles called *soil colloids.* While the clay particle is very small (Figure 1, page 1), the soil colloid is only half that size.

There are two kinds of colloids in soils: *inorganic colloids* (mineral or clay) and *organic colloids* (humus). The inorganic mineral clay colloids have been photographed with an electron microscope and found to be flat, plate-like particles like flakes of mica. The exact shape of the organic humus colloid is unknown, but both types carry a negative electrical charge.

How Colloids Attract Nutrients

Since negative attracts positive, the negatively charged soil colloid (−) attracts the positively charged ions (+) which carry the plant food elements. Hydrogen ions are also attracted to the soil colloid (Figure 10).

There are two groups of plant food elements in the soil. The first group includes the elements calcium, magnesium, potassium and sodium, which all carry a positive charge and are called *base elements.* Each element is in the form of an ion with one or more positive charges. Calcium, for example, has two positive charges and is written Ca^{++}. Magnesium is written Mg^{++}. Potassium is written as K^+.

As a magnet attracts pieces of iron, these ions are attracted to a soil colloid when they get near one, and this is known as *adsorption.* Now a curious thing happens. Nitrogen is not a base element, but when it runs around with hydrogen it acts like one by taking on a positive charge. This is written NH_4^+ (Figure 10). The other form of nitrogen upon which plant roots feed is NO_3^-; scientists think this form isn't adsorbed, but that it moves around in the soil solution instead.

The second group of plant food elements in the soil are also in the form of ions, but they carry a negative charge. One is phosphorus, written

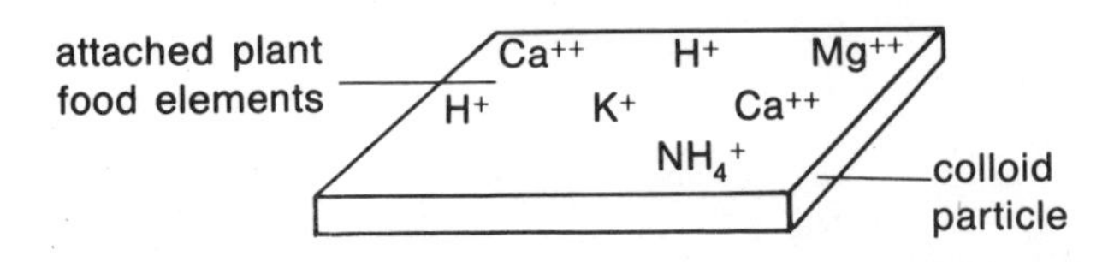

Figure 10. Soil colloid (about 2500 times actual size).

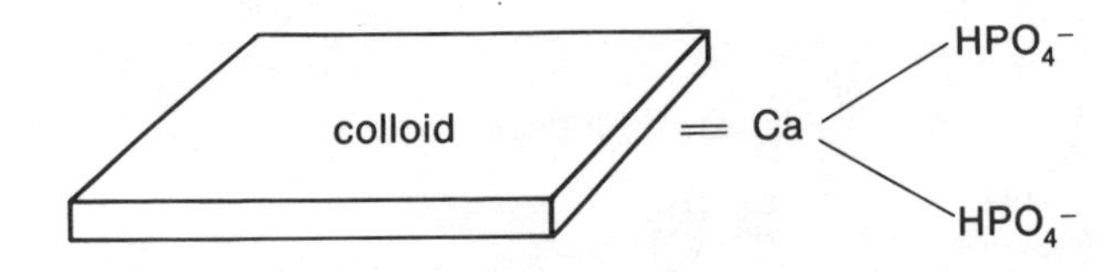

Figure 11. Absorption of phosphorus on a soil colloid with the help of a calcium ion.

HPO_4^{-2}. Can it be adsorbed and held by soil colloids? Recent research indicates that it can be held by hooking onto base ions like calcium already attached to the colloid (Figure 11). Sulfur is probably held this way also. Chemically speaking, nitrogen phosphorus and sulfur are the only major plant food elements in the soil that are not metals. Elemental nitrogen is a gas, elemental sulfur is a yellow powder, and elemental phosphorus is a soft, white, waxy solid.

You might think of soil colloids as a comb that has been pulled through your hair and is so charged with static electricity that it will pick up small pieces of paper and hold them.

How Roots Feed from the Soil

Plant roots (root hairs) penetrate the pore spaces of the soil like the tentacles of an octopus. Within these pores is the soil solution with its dissolved plant food elements (Figure 12). By complex biochemical processes called active absorption, the plant food elements pass from the soil solution into the root, from the root into the stem, and from the stem into the leaves for photosynthesis.

But what about the plant food elements electrically attached to soil colloids? How does the plant root get at them and pry them loose? By two means: (1) with help from soil organic matter, and (2) by direct biochemical action of the plant root hairs.

Organic matter. Organic matter decomposes and gives off carbon dioxide gas, which combines with water to form carbonic acid. Carbonic acid is rich in hydrogen, and when it touches a soil colloid loaded with plant food elements the carbonic acid

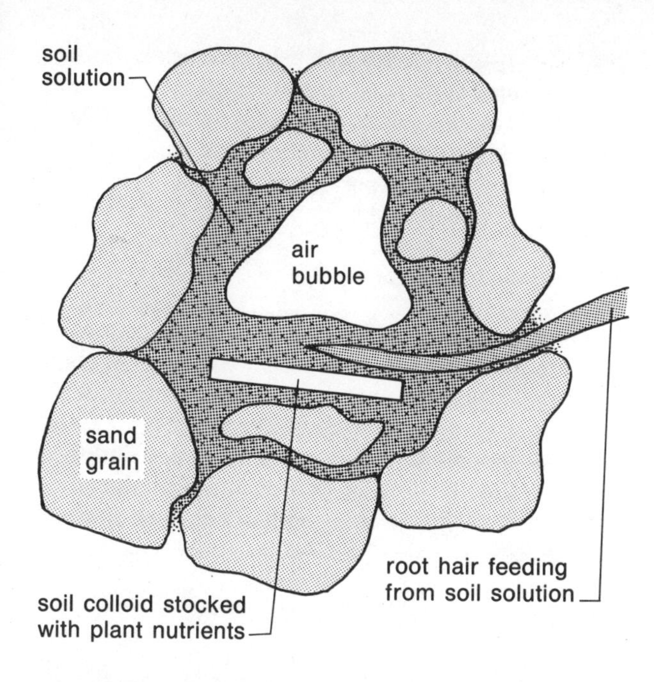

Figure 12. Plant root feeding from a soil solution containing colloid particles.

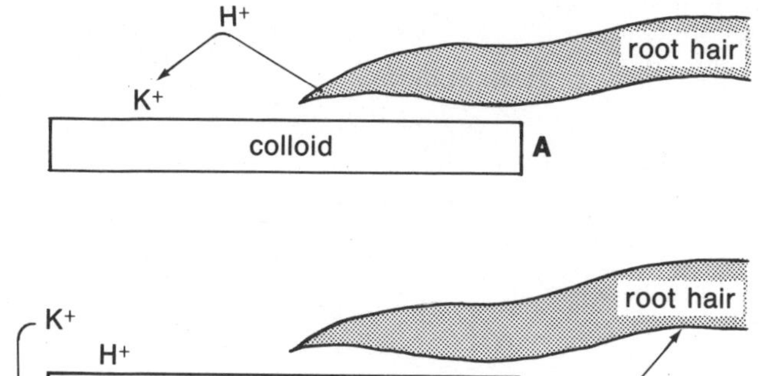

Figure 13. Root hair of a plant feeding directly from a colloid (7,000 times actual size). A—beginning of cycle as hydrogen ion is secreted from plant root and displaces potassium ion on soil colloid. B—a microsecond later the displaced ion goes into the soil solution, and then into the root hair.

gives up hydrogen, which moves over to the colloid to knock off elements like potassium or calcium. The element that is knocked off goes into the soil solution, where the root hair is feeding (Figure 13).

When this organic matter is decomposing and the plant food elements are getting knocked loose from the soil colloids into the soil solution, what happens if there are not plant roots to absorb them? If no heavy rains or heavy irrigations are applied, the released elements just enrich the soil solution. But if you irrigate heavily or a heavy rain comes and the soil is sandy loam or loam, the excess water will move down into the pore spaces and give your soil a good washing, moving the rich soil solution out of your soil and into the drainage waters that eventually find their way into streams. The loss of plant food elements this way is known as *leaching.*

In arid, low-rainfall areas the elements entering the soil solution are not subject to uncontrolled rainfall and so are not as likely to be leached. But they tend to accumulate, making these soils very high in salt and mineral content, and the soil pH alkaline.

In humid areas where average annual rainfall is 35 inches or more, leaching moves the base elements out of the soil, and hydrogen accumulates on the soil colloids to make them acidic.

Feeding by direct root action. The second way plant roots feed is by direct, biochemical action of the root hairs. Hydrogen is the key that unlocks the pantry door where the goodies are kept. The hydrogen ion has a unique ability to release plant food elements from the electrical grip of the colloid.

If there's no organic matter around to furnish the hydrogen, where will it come from? From the root hair itself—the root hair releases hydrogen ions which knock plant food elements off the colloids into the soil solution for absorption by the root hairs.

For all practical purposes, the plant root feeds directly from the soil colloid as well as from the soil solution. Proof of this is in the arid West where many soils have practically no organic matter yet yield bumper crops. The plant roots furnish hydrogen to exchange for the elements attached to colloids in the soil.

Sand Culture and Plant Feeding

Sand culture is commonly used in greenhouses. A layer of sand is kept moistened with a nutrient solution (water and fertilizer). Fabulous quality crops are grown this way—and no colloids are present. All plant root feeding is done via the soil solution, which is completely under the control of the gardener and where no irrigation or rainfall can leach out plant foods.

Plant Food Elements

Thirteen elements essential to the proper growth of plants must be supplied by the *soil.* Two other

Table 7
Essential Plant Food Elements

Symbol	Chemical Element	
C	carbon (from carbon dioxide in atmosphere)	
H	hydrogen (from soil water)	
O	oxygen (from soil water)	
N	nitrogen	*primary elements*
P	phosphorus	
K	potassium	
Ca	calcium	*secondary elements*
Mg	magnesium	
S	sulfur	
Fe	iron	*trace elements or micro-nutrients*
Mn	manganese	
Zn	zinc	
Cu	copper	
B	boron	
Mo	molybdenum	
Cl	chlorine	

essential elements are supplied by water (hydrogen and oxygen) and one is supplied by the atmosphere (carbon—as carbon dioxide). See Table 7.

These 13 elements (essential nutrients) are usually found in naturally fertile soils. But they may be supplied by fertilizers (see the section on fertilizers, pages 40-57).

How do we know these 13 elements are essential and others are not? Scientists have made many careful tests where one element at a time was left out of a nutrient solution used to feed plants. Plants kept right on growing when some of the elements were left out; when some others were omitted, the plants got sick.

One essential element—chlorine—is never listed on bags of fertilizer. This is because it is always present in other materials, so you don't have to deliberately add it to the soil. It gets there incidentally.

As a plant grows it builds up its tissues mainly from carbon, hydrogen and oxygen (95 percent) and uses only a small amount of mineral elements (5 percent). When a plant "gains weight," it uses only a little mineral matter doing it. Most of the weight of green plants is water (about 75 percent).

Primary Elements

Among the "Magic 13" essential elements are the "Big 3," the primary elements nitrogen, phosphorus, and potassium. Nearly 60 percent of a plant's *mineral* intake consists of these three elements. When you look at a bag of fertilizer, you'll see three numbers, such as 5-10-5. These numbers refer to the "Big 3" because plants use so much of them. The 5-10-5 refers to the percentages of each of the three primary elements in the bag. (Actually, phosphorus and potassium are in compound form. See pages 40-41.)

Secondary Elements

After the "Big 3" are three secondary elements: calcium, magnesium, and sulfur. About 39 percent of a plant's mineral intake consists of these three.

If you add the amounts of the six primary and secondary elements used by plants, you'll find that the plants get 99 percent of their mineral diet from six elements. The other 1 percent consists of trace elements.

Trace Elements or Micro-Nutrients

Since only 1 percent of a plant's intake of mineral elements comes from trace elements, does this mean they are not important? By no means.

There are seven trace elements essential to plants. We call them "trace" elements because when you analyze a plant for their presence you find only a trace or tiny amount. They are minuscule in amount but not in importance. If a single one of them is missing, a plant fails to grow properly and may die.

The seven trace elements are iron, manganese, zinc, copper, boron, molybdenum, and chlorine.

Sources of the Primary Elements

Nitrogen. As an element, nitrogen is a colorless, odorless, tasteless gas in the air we breathe. About 80 percent of our atmosphere is nitrogen gas and about 20 percent is oxygen gas.

In the soil, nearly all the essential elements occur not as elements but in combination with other elements, usually oxygen. Nitrogen will be found as protein in organic matter tied up with carbon, oxygen, sulfur, and phosphorus. As it is decomposed by microbes (soil microorganisms) the protein is broken apart, and finally ammonium ions (NH_4^+) or nitrate ions (NO_3^+) appear. The plant cannot feed directly on the organic matter nor on the protein in it, but once it is broken down (mineralized) into the two ions above, the plant begins to feed.

When you apply salt-type (or "commercial") fertilizer, you short-circuit the process above, where nitrogen is biologically unlocked. Instead, you're applying "pre-digested" plant foods such as NH_4 and NO_3.

Nitrogen is the only essential element from the soil that does not originate there. All the others exist in rocks. During soil-building processes where over millenia, natural forces grind rocks into powder, these essential mineral elements are released and become part of the soil.

Phosphorus. Like nitrogen, phosphorus is a part of the protein present in organic matter and is broken down for plant use the same way nitrogen is. But phosphorus is mined out of the ground like coal. It is found as a mineral deposit combined with calcium, fluorine, and oxygen.

Phosphorus content on a fertilizer bag is indicated by the middle number, for example the "10" on a bag of 5-10-5. Actually, the phosphorus is combined with oxygen to form phosphate (P_2O_5). The "10" means 10 percent phosphate, not 10 percent phosphorus.

Potassium. Potassium does not occur in nature. It is man-made by an electrical process. Potassium is combined with oxygen to form potash (K_2O), and this is the form in which it is used by plants.

Potassium content in a fertilizer is indicated by the third number—the last "5" on a bag of 5-10-5, for example. As with phosphorus, this number means percent potash present, not percent potassium.

Trace Elements for Human Nutrition

If you're interested in growing your own food and getting the most nutrition out of it, you may want to add four other elements to your soil: cobalt, iodine, fluorine, and sodium. These four are trace elements needed in tiny amounts in the human diet. One way to add them is to fertilize with seaweed, which is rich in iodine and other trace elements.

Essential Soil Elements

Nitrogen

Function. Nitrogen is used for amino acids, the building blocks of protein. It is also found in the chlorophyll molecule, which promotes dark green leaf color and gives plants their photosynthetic capacity.

Deficiency symptoms. Slow growth. Stunted plants. Leaves yellowish-green. "Burn" on tips and margins of older leaves.

Excess. Excessive growth. Delayed maturity. Weak stems. Poor quality. Low resistance to diseases and insects. May induce copper deficiency.

Phosphorus

Function. Stimulates early growth. Promotes root formation. Hastens maturity. Promotes flowering, fruiting, seed formation. Counteracts adverse effects of too much nitrogen. Promotes quality and resistance to diseases.

Deficiency symptoms. Slow growth. Stunted plants. Purplish color of leaves (on some plants). Dark green leaf color with tips dying. Delayed maturity of grain, fruit, seed.

Excess. May induce zinc, iron, and copper deficiencies.

Potassium

Function. Promotes starch and sugar formation. Increases disease resistance. Promotes strong root system. Balances adverse effects of too much nitrogen or phosphorus. Promotes root and tuber formation.

Deficiency symptoms. Tip and margin burn of older leaves. Weak stems, small fruit, shriveled seeds, slow growth.

Excess. Overconsumption—plants take up more than they really need. May induce manganese deficiency.

Calcium

Function. Promotes formation of new cells.

Deficiency symptoms. Death of terminal buds. Abnormally green foliage. Premature shedding of blossoms and buds. Weak stems.

Excess. May induce trace element deficiency by raising soil pH. May induce boron uptake.

Magnesium

Function. Helps build chlorophyll. Activates plant enzymes.

Deficiency symptoms. Chlorosis (yellowing) of veins in older leaves. Leaves turn yellow along margins, with inner green "tree-shaped" area. Leaves may curl upward at edges.

Excess. Same as calcium.

Sulfur

Function. Helps build plant protein. Gives flavor to certain crops.

Deficiency symptoms. Younger leaves yellowish-green. Small, weak plants. Delayed maturity, slow growth.

Excess. Can interfere with uptake of certain elements by lowering soil pH.

Iron

Function. Helps build chlorophyll.

Deficiency symptoms. Leaf veins remain green while rest of leaf turns yellow. Dieback of twigs.

Excess. May reduce uptake of manganese.

Zinc

Function. Growth regulator.

Deficiency symptoms. Short internodes of stems. Leaves form rosette at tips of branches. Mottled leaves. Reduces fruit buds.

Excess. May reduce uptake of iron and manganese.

Copper

Function. Enzyme activator. Promotes vitamin A formation.

Deficiency symptoms. Stunted plants. Dieback of twig tips.

Manganese

Function. Aids in chlorophyll formation.

Deficiency symptoms. Yellowing of leaves.

Boron

Function. Regulates carbohydrate formation.

Deficiency symptoms. Terminal growth dies. Leaves thickened, curled, chlorotic. Soft spots in fruits or tubers. Reduced flowering or pollination. Poor fruit or seed set.

Molybdenum

Function. Aids protein formation.

Deficiency symptoms. Stunting. Burn of leaf edges; cupping, or rolling, of leaves.

Chlorine

Function. Aids photosynthesis.

Deficiency symptoms. Rarely noted.

Soil Organic Matter

Organic matter is the nonliving remains of plants and animals. It can be the nonliving bodies of plants (stalks) or the non-living bodies of animals such as earthworms or, with higher animals, parts shed such as hair and feathers or waste matter excreted.

Raw organic matter is the remains of dead plants or animals before decomposition. It is usually made up of old plant stalks or leaves or fresh manure. It is bulky, and if put in a pile and kept moist, it gets hot. This is because the decomposition of raw organic matter, as in a compost pile, is a form of slow burning (oxidation). Like other forms of oxidation, such as the rapid form when you burn wood in a fireplace, heat is always given off. As the organic matter gradually decays it shrinks in bulk and becomes the dark, earthy-smelling mass we call compost.

Decomposed organic matter such as compost is also called *humus.* Humus has considerable "staying power" in the soil: If you were to fill a compost bin full of humus (finished compost) and keep it moist and warm, the pile would remain more or less intact for a long time and not shrink very much in size. The same bin filled with raw organic matter would rapidly decrease in size and end up as half its original mass.

Microbes in the soil feed upon raw organic matter and break it down into humus. These microbes work best when the raw organic matter contains about 30 parts of carbon to 1 part of nitrogen. They use the carbon for energy, as we burn the carbon in coal or wood for heat energy. They use the nitrogen to build up body tissue much as a youngster must have protein to grow.

When the microbes are all done, the 30-to-1 raw organic matter will be changed into humus with a 10-to-1 carbon-to-nitrogen ratio.

Raw sawdust has an unbelievably high carbon-to-nitrogen ratio of 511-to-1, or 511 parts of carbon to 1 part of nitrogen. Imagine the frustration of the soil microbes when they find raw sawdust for dinner! But like most living things, they're always hungry and aren't about to let any food get away, no matter how poor. So, rather than go hungry, they adapt to the situation. They *must* have extra nitrogen to work; can you guess where they get it? Right out of your soil.

So if you have any plants growing in a garden soil mixed with raw sawdust, watch out. Their leaves will begin to turn yellow as they starve for nitrogen. Why? Because the microbes have a greater ability to extract nitrogen from the soil than do the roots of your garden plants. Microbes always win the battle. We say the microbes have "robbed the soil of nitrogen." To overcome this, you have to add supplemental nitrogen to the soil (see pages 50-51).

How Plants Feed on Organic Matter

Plants cannot feed on organic matter directly. They must wait until it decomposes into simpler substances. Scientists call this process *mineralization.* Nitrogen, phosphorus, and sulfur, for example, are tied up in organic matter as proteins. The mineral plant foods are also locked up. And the cement keeping the plant foods locked up in the organic matter is carbon.

When microbes attack organic matter, it begins coming apart as the carbon is pulled out and combined with oxygen to form carbon dioxide gas. This allows the plant foods, like nitrogen and phos-

Sawdust as a Soil Amendment

Sawdust is so easily available and so rich a source of organic matter that it would be a crime not to make use of it. But you can only use raw sawdust if you balance it by adding nitrogen. (Or, you can simply apply it to the soil surface as a mulch, and not work it in at all.)

If you want to work it in, compost it first. Try the following procedure:

1. Mix equal parts of sawdust, manure or sewage sludge, and sand.
2. Spread 6 to 8 inches deep in a conveniently located area.
3. Turn every 3 or 4 weeks during warm weather.
4. Water during dry periods.
5. After about 3 stirrings and 3 months, when the material has turned a dark color, it should be ready for use.

Sawdust and Poultry Manure

Since sawdust is notoriously low in nitrogen (0.2 percent) and poultry manure is high in nitrogen (6 percent), you can mix these two and convert piles of useless sawdust into valuable organic fertilizer.

A mixture of 4 parts sawdust and 1 part manure (dry weight) should be about right. Keep moist and aerated by turning occasionally. After a few weeks you'll have the 1.2 to 1.5 percent nitrogen content—or a 30:1 carbon-to-nitrogen ratio—necessary for efficient decomposition by soil microbes.

Sawdust and Chemical Fertilizer

If you want to mix raw sawdust directly into your soil without composting, you'll need to add about 1¼ lbs. of ammonium phosphate (16-20-0) per bushel of sawdust (about 1¼ pints of ammonium phosphate per 10 gallons of sawdust).

Sawdust is also low in phosphorus; hence the advice to use a fertilizer rich in both nitrogen and phosphorus (16-20-0).

Aged Sawdust

This is simply old sawdust that has been around awhile. It will be a darker color and is probably the best type of untreated sawdust to use. If placed in a shallow layer, say 8 inches deep, and kept moist, it is possible that the free-living, nitrogen-fixing bacteria, *Azotobacter,* would appear, live in the sawdust and capture atmospheric nitrogen free of charge. Research should be done on this. Perhaps a culture of *Azotobacter* could be sprayed on sawdust to get the reaction started and phosphate fertilizer added to supply phosphorus.

phorus, to become mineralized. Now plants can absorb these elements combined with oxygen or hydrogen in the form of ions.

Humus—an Organic Colloid

Remember what we said about colloids on pages 21-22? A soil rich in colloids meets one of the requirements of a fertile soil. But if the colloids are inorganic (as the clay colloids in clay soils are), the soil will be packy, sticky when wet, and difficult to manage. What can you do?

You can fill your soil with organic colloids found in humus. We said before that soil colloids are like a big pantry for storing plant food in the soil. Humus colloids make up a king-size pantry since they can hold up to 30 times as much plant food as clay colloids. And they do this without the hard-to-live-with behavior of clay.

Humus also has a tremendous capacity to absorb water. This is of special importance in sandy soils, which hold little water. Increasing the organic matter content of sand to 5 or 10 percent will greatly increase its ability to catch and hold water.

Phosphorus and Organic Matter

Plants use large amounts of phosphorus, but most of the phosphorus present in soils is unavailable to them; and when inorganic chemical fertilizers are applied, the phosphorus often reacts with the soil again and becomes unavailable. Studies have shown that when inorganic phosphorus fertilizers are applied to soils, plants recover only 10 to 20 percent of it. Yet, other studies show that the presence of microbes and organic matter in the soil greatly increases phosphorus availability. So, to be sure the plants get plenty of phosphorus, keep the soil high in organic matter that is also organic *fertilizer* (peat moss is organic matter, but it has no fertilizing value).

Kinds of Organic Matter Compared

We have already discussed sawdust and touched briefly on manure. All forms of organic matter are soil amendments, that is, they improve soil structure. They help make tight soils loose and loose soils tighter.

How often do you need to apply organic matter to keep your soil in good shape? This depends on how long it (the organic matter) will persist in the soil, and persistence depends on the *lignin* content. Lignin is a component of organic matter that is so

Table 8
Percent Composition of Some Kinds of Organic Matter

	Lignin	Nitrogen	Phosphorus	Potassium
Activated sewage sludge	High	5.5	5.5	0.4
Farm manure	—	1.2	0.6	1.2
Sphagnum peat	18-19	1.0	0.05	0.04
Sawdust (general)	18-28	0.2	0.1	0.2
Sawdust (Western Yellow Pine)	27	—	—	—
Sawdust (Hickory)	23	—	—	—
Reed and sedge peat	35-49	1.9	0.1	0.05

tough it resists decay to the bitter end and so persists in the soil a long time. Some kinds of organic matter stay in the soil much longer because they have a higher lignin content (Table 8). No data are available on shredded bark, but it should also be high in lignin.

Now let's look briefly at the common kinds of organic matter and their suitability for improving garden soils.

Peat moss. Peat moss (peat) is the remains of swamp plants. The largest deposits are in Minnesota. Though it is a good soil amendment, peat moss has no fertilizer value.

The light-brown type consists of the remains of grasses, reeds, or sedges and is usually acid in reaction. (Baled peat moss is this type.) It is fibrous and has a very high water-holding capacity; it contains about 1 percent nitrogen but is low in phosphorus and potassium.

The other type of peat moss is brown to black and granular (non-fibrous); it ranges from acid to slightly alkaline in reaction.

Dry peat moss should be moistened before adding to potting soils.

Sphagnum moss. "Sphagnum peat moss" and "sphagnum moss" are the same thing. It can be either shredded (ground) for use in making potting soil or fibrous (with long fibers) for use in making hanging baskets for houseplants.

Sphagnum is light brown in color, relatively sterile, and light in weight. It holds water like a sponge—it will absorb 10 to 20 times its weight in water. A pint of sphagnum moss will weigh about 2 oz. when dry but will absorb 1 to 2 pints of water and change from the original 2 oz. dry to a pound or two when fully moistened.

Sphagnum moss is very acidic (pH 3.5), so you must add limestone when using it in a potting soil (unless you're growing acid-loving plants). It also contains a natural fungus-inhibiting substance helpful in preventing damping-off disease of young seedlings.

Sphagnum moss is the king of the peat moss group. It is a main ingredient of the well-known synthetic soil, peatlite, which is half sphagnum and half vermiculite.

Leaf mold. The leaves of deciduous trees such as oak and maple may be collected and turned into a valuable soil amendment. The process is to build a wire enclosure of 1-inch poultry netting 36 inches tall in which to place layers of leaves.

The size of the enclosure depends on how many leaves you have. A size of 10 feet by 10 feet is a good start. Wood posts 2 x 2 inches in size and 4 feet long are about right. Dig holes 12 inches deep and paint the bottom 14 inches with penta to prevent rot. Place the four posts at the four corners, and tamp soil around the posts well to make them steady.

To load the enclosure, apply leaves in 12-inch layers and wet each layer with a hose. Then climb in and tread the leaves down, apply another layer, and repeat the process. Build other layers to within 6 inches of the top. Cap off the top with 3 inches of loose leaves and water down good. If you have some compost, sprinkle a little on top of each layer of leaves.

Water the leaf bin occasionally to keep it from drying out. In about a year you should be able to rake off the top layer and find good, rich, dark-colored, earthy smelling leaf mold ready to use.

Shredded bark. Bark comes in fine, medium and coarse grinds. The coarse and medium grinds are used for decorative mulching, especially in dry, windy areas where wind tends to blow mulch away. The fine grind of shredded bark is used as a soil amendment and in potting soils. Bark is higher in lignin than most other forms of organic matter so it lasts longer.

Sewage sludge. Sewage sludge is basically human manure. It is the solid material that ends up at the bottom of the digestors down at the sewage plant. Like steer manure, it is a low analysis fertilizer and is being tested for use as a fertilizer as energy costs go up. Fossil fuel is widely used in the manufacture of salt-type, highly concentrated fertilizers, and many energy conservation groups are researching ways to recycle organic fertilizers now being hauled to the dump.

The two main problems with sewage are (1) disease-producing bacteria, and (2) pollutants from sewage plants in industrial areas.

Table 9
Composition of Typical Sewage Sludge from Industrial Areas (Dry Weight Basis)

Element	%	Parts per Million	lbs./ton
Nitrogen	3.5-6.4	—	70-130
Phosphorus	1.8-8.7	—	36-170
Potassium	0.24-0.84	—	4.8-17
Zinc	0.24	2400	4.8
Copper	0.09	900	1.8
Lead	0.04	400	0.8
Nickel	0.022	220	0.44
Cadmium	0.006	60	0.12
Mercury	0.0015	15	0.03

Disease-producing bacteria and viruses (pathogens) are common even in digested sewage sludge. Some state health departments do not at present recommend the use of sewage sludge on land where any food crop to be eaten raw is grown (lettuce, radishes, etc.).

But EPA researchers believe that sewage sludge is all right if applied and worked into the soil in the fall on land to be planted next spring. After soil application, disease-producing bacteria are eliminated rather quickly. During summer conditions, 99.9 percent of fecal coliforms and streptococcus are killed within 40 days after soil application.

Industrial pollutants. Practically every kind of waste there is goes into the sewer. In industrial areas, heavy metals like copper, lead, zinc, nickel, and mercury may be dumped in the sewer and end up contaminating sewage sludge (Table 9).

The conclusion reached after 3 years of tests in Pennsylvania by Dr. D.E. Baker is that municipal sewage sludge from *industrial* areas contains too much cadmium for safe use as a fertilizer. Some sewage sludges were found to contain over 300 ppm of cadmium; 60 ppm is more typical (Table 9). Application of sewage sludge has a "poisoning" effect upon the soil. "The effects are irreversible," says Dr. Baker, "only 2 lbs. of cadmium per acre could maintain cadmium content in crops at 5 to 10 times the normal levels for 100 years."*

So, if you're in an industrial area, or if the sewage sludge comes from an industrial area, don't use it.

Gin trash. In or near farming areas where cotton is grown, gardeners often cast hungry eyes toward the huge piles of gin trash that could improve their soils.

*Hort Science, Vol. 11, No. 5, p. 451.

Beware! All that glitters is not gold. Gin trash is a form of organic matter and is free for the hauling, but it is *polluted.*

Gin trash is a by-product of cotton processing. Mechanical pickers pick the cotton and haul it to the gin where the cotton lint and seed are removed, leaving cotton leaves, twigs, and burrs as gin trash.

But before the cotton was picked it was very likely sprayed with insecticides such as chlorinated hydrocarbons and harvest-aid chemicals (defoliants such as arsenic acid) which show up as residues in the gin trash. A study in Texas showed several persistent chemicals, particularly arsenic acid, in gin trash.

For these reasons it is not advisable at this time to use gin trash as a soil amendment.

Lawn clippings. Lucky you are if you have a good non-Bermuda grass lawn that needs mowing often. Grass clippings are an excellent source of organic matter if free from noxious weed seeds. Spread them out in a 2-inch layer to partially dry before using. This will prevent a soggy, smelly mass (and mess).

Bermuda grass lawns, if under high-level management, are a good source of clippings too. High-level management means keeping the grass growing so rapidly it forgets to set so many seed heads, or if it does, you mow before the seed heads have time to set seeds. This cuts down on pollen (and allergy trouble), gives you a more beautiful lawn, and is a source of organic matter rich in nitrogen.

Seaweed. Ocean vegetation is an excellent source of organic matter and, like manure, it is also a good, low-analysis fertilizer. If you live near a coastline, harvest this valuable source of organic matter. It blows up on the beach from time to time and may be loaded onto a trailer. Every time you go to the beach, take along an empty burlap bag and bring it home full of seaweed. Even a little will help your garden soil.

Seaweed is superior to manure and probably the best form of organic matter you can get because it contains every known mineral element—those known to be essential and those not yet proven to be so.

Cottonseed hulls. When cottonseed is milled, the contents of the seed are removed, leaving the outer coating, or hull. These are known as cottonseed hulls. This material is lightweight, relatively inexpensive, and is suitable as a soil amendment.

Manure. Manure is an excellent form of organic matter and the most common. It is also a low analysis, organic fertilizer. All forms of manure are good—steer, dairy, horse, rabbit, and poultry.

But use feedlot manure with caution—it has a high table salt content (sodium chloride). Steers in a feedlot are nearly always force-fed salt. Most feedlot manure will contain 10,000 ppm or more of sodium, whereas 900 ppm or less is normal. The 10,000 ppm of sodium is too high for many crops. Manure is discussed in detail on page 48.

Disease Control with Organic Matter

Some soils, especially those high in organic matter, allow little or no soil-borne plant disease to develop, according to a study made in Australia by Dr. R.J. Cook of Washington State University.* Dr. Cook says large amounts of organic matter in the soil may lead to production of ethylene gas, which he thinks suppresses soil-borne diseases such as root rot.

Hauling Problems

The main problem with organic matter is material-handling. It is so bulky that it is often troublesome to load, haul, and unload.

A pick-up truck with side rails or an automobile and trailer with side boards is essential for hauling organic material to your garden.

For sources of organic matter like manure, check the classified ads of your newspaper. Often, a farmer will haul a load of manure to your backyard at a reasonable cost.

Spoiled hay can often be had for the hauling away or at a very nominal cost. To find a source of spoiled hay, take a Sunday drive through a farming area and look for old hay stacks fallen over. Make inquiries to see if you can haul some of it home. Watch out for and avoid Bermuda grass or Johnson grass hay.

Inorganic Soil Amendments

Vermiculite. Vermiculite consists of small particles of whitish, corky, very light material. It is made from thin sheets of mica with entrapped water. When heated to high temperatures, it pops like popcorn into small, porous, spongy kernels.

The best kind of vermiculite to use for gardening is # 2 "Horticultural Grade," with particles about

***Organic Gardening*, March 1977, p. 4.

1/10 inch in size. It will pay you to check this point carefully and accept no substitutes. Many times, all the vermiculite the garden center has on-hand will be a finer grind. Do not buy it; ask them to order the coarser grade for you or look elsewhere. Take a ruler with you and measure some of the particles; if they're 1/10 to 1/8 inch across, you have the right size.

The reason you *must* use the horticultural grade in gardening is that vermiculite is also used for insulation, and this type is waxy and will not absorb water.

Vermiculite is mostly silicon, the main component of sand. But vermiculite also contains considerable amounts of the plant food elements magnesium and potassium.

Vermiculite is superior to perlite not only because it is a source of plant food elements but also because it has a great ability to capture and hold other elements added to the soil.

Vermiculite is *not* organic matter. It is mineral, or inorganic, matter with many of the traits of organic matter. It makes soil spongy, increases water-holding capacity and ion exchange capacity (see pages 21-22) just like organic matter. But remember, it's made of exploded mica. About the only problem with vermiculite is its expense compared to other soil amendments. But then, you usually get what you pay for.

Perlite. Perlite, too, consists of particles of white, spongy material. It is also not organic matter, but it will give good structure and water-holding capacity to potting soils. It originates as a volcanic rock and is "popped" by heat treatment similar to the way vermiculite is made.

Perlite will hold 3 to 4 times its weight in water. It is sterile, has a neutral pH, no plant food content, and has no exchange capacity; thus it cannot store, hold, and later release foods for plant growth. It is relatively inexpensive.

All About Compost

If you garden at all, you already have the beginning of a fertilizer factory in your own yard. The raw material is on hand: the old, dead stalks from your garden. The end product, compost, or humus—that dark, spongy, earthy-smelling material—is the world's finest organic fertilizer.

Good compost is not an accident—it is the product of a scientific process. The best way to make compost is in bins. There are two good reasons for this: First, without bins a compost pile won't heat up properly and stay that way long enough to work unless it is huge, like 4 feet high by 8 feet wide by 8 feet long. Most gardeners just don't have that much material or space. The second reason to use a bin is tidiness. Most home gardeners have landscaping features in their backyards, and nothing can become messier than a pile of compost lying about.

In a properly made compost bin 3 to 4 feet square, the heap will heat up to about 158 degrees by the third day and cool to 110 degrees, staying this way for about a week. This not only converts unsightly old stalks, etc. to rich humus but also kills any pathogenic bacteria that might be in the heap. This is why even sewage can be composted and turned into rich organic fertilizer if the composting is done *properly*. (For details, see page 29.)

The chemical composition of compost varies widely. A gallon of screened, moist, homemade compost weighs about 5.25 lbs. When dried, this same gallon of compost weighs about 3.5 lbs., about the same as an equal volume of manure.

Making Something from Nothing

Waste materials from garden, flower bed, and kitchen, can be placed in a pile and burned. The burning is a very rapid form of oxidation whereby the carbon in the organic matter is converted by intense heat to carbon dioxide and a small pile of mineral matter (ashes). The ashes make good fertilizer, but much is wasted because you have lost the humus that was in the pile of organic matter.

So, it's far better to convert the pile of waste material into humus. This is easily done by composting, a very slow form of oxidation (burning). Compost "burns" at a temperature of about 110 degrees for a week, converting 200 gallons of waste material (1 cubic yard) to about 100 gallons of compost (humus).

What Is Humus?

Humus is defined as "a complex aggregate of amorphous substances resulting from microbiological activity in the breakdown of plant and animal residues." Humus is a dark brown or black soil-like substance with an earthy smell; it is insoluble in water (it won't leach out of the soil); it has a high ability to attract, hold, and give up nutrients to plants; and it continues to slowly "burn" or oxidize in the soil, giving off carbon dioxide, water, and plant nutrients. Humus and compost are one and the same thing.

Composition of Humus

Humus varies widely in chemical analysis, but a typical analysis of dried compost is shown below as percentage by weight:

Organic matter Composition (%)	*N*	*P*	*K*	*Ca*	*Carbon*	*Ash*
25-50	0.4-3.5	0.3-3.5	0.5-1.8	1.5-7	8-50	20-65

N = nitrogen; P = phosphorus (as phosphate, P_2O_5); K = potassium (as potash, K_2O); Ca = calcium (as calcium oxide, CaO).

Microbes Make Humus if Conditions Are Right

Microbes (microorganisms), in the form of bacteria and fungi, do the "pick-and-shovel" work in breaking down the compost heap to form compost. They must have oxygen and moisture to do this work. They are thermophilic (cause large amounts of heat to be given off while at work). They require about 30 parts of carbon (C) to one part of nitrogen (N), so organic matter with a carbon-to-nitrogen ratio of about 30:1 is most favorable. If the C:N ratio of the compost material is less than this (young grass clippings, for example, have a ratio of about 12:1), there is more N than the microbes can use; this is why you smell ammonia around hen houses and piles of grass clippings—the extra nitrogen is coming off as ammonia gas (NH_3), which is a waste of nitrogen.

Therefore, some high-carbon (woody) material such as straw should be mixed in. If, on the other hand, too high a C:N ratio exists (wheat straw, for example, with 128:1), there is not enough nitrogen, so the microbes work very slowly and composting takes a long time. You have some latitude in this ratio, however; just try to balance nitrogenous materials like grass, leaves, and manure with carbonaceous materials such as straw, sawdust or other "woody" things. See Table 10, page 33.

The composting process can be either aerobic (requiring oxygen) or anaerobic (requiring no oxygen), or both. The aerobic type is faster.

The Art of Making Compost

When to Make Compost

A good time to make compost is any time you can get (or have accumulated) enough material. If you are using a bin 3 to 4 feet square by 3 to 4 feet high, it doesn't take too much material. The best times are (1) at the end of the growing season before the weather gets too cold which is (usually August to October), (2) in early spring if you missed the fall season, or (3) on a continuous basis where you have small amounts of material daily or weekly such as non-mushy garbage or other refuse.

The Best Place

A bin with insulating walls is ideal. If you make it 5 feet wide by 5 feet long by 4 feet high, it will be small enough to fill easily, and the insulating walls will allow it to heat up properly. Use 2 x 6s or concrete blocks for the walls. The front of the bin should be removable for easy filling and harvesting of the compost.

Three bins are best—one for holding accumulating wastes, one for a compost heap in the process of composting, and one as an empty bin for turning or storing finished compost. If three bins are too much, try two for faster decomposition.

Wire-covered frames, though not as good, can also be used with hook-and-eye fasteners at the top four corners and wood stakes at the bottom corners to hold the four walls in place.

Any service area in your backyard, preferably close to your vegetable garden, is a good place for your bins. If they're of masonry, footings are good to use but no bottom is necessary. If you use concrete blocks for your bins, trenches 6 inches deep filled with 3 inches of packed and leveled sand will do as a foundation.

A deluxe set-up is a concrete slab at least 6 feet by 6 feet, handy for screening compost, mixing soil, etc.

Homemade compost bin of brick walls and slatted movable board front. Note pitchfork for turning and sprinkler for watering during dry periods. This bin is 4' × 4' × 3' high.

Composting Facts

- The rate of decomposition of a compost heap depends on the nitrogen content of the materials in the heap. The more nitrogen, the more rapid the composting process.

- History of a compost heap: In a properly made heap with nitrogen content of 0.8-1.7 percent, the temperature rises to about 120 degrees Fahrenheit after 1 day and to 158-167 degrees in 3 days. The 158-167 degrees will persist for about a day then drop to about 110 degrees for about a week if the heap has good aeration. A compost heap working properly has the following characteristics: (1) After building, it shows a rapid rise in temperature, followed by a leveling off and slow decline in temperature. (2) no smelly odor, and (3) a progressively darkening color.

- Grinding of materials (and mixing) of materials for the compost are necessary for rapid aerobic composting. Size of grind should be 1½ inches or less.

- Without bins, the dimensions of a heap should be: 4-6 feet high x 10 feet wide x 10 feet or more long.

- Proper moisture content is 40-65 percent.

- Finished compost has (1) a dark gray or dark brown color, (2) neutral, slightly musty or earthy odor, and (3) a carbon-to-nitrogen ratio of about 20:1 or less.

Size of the Compost Pile

A properly built compost pile must be well aerated, have the right moisture content, and retain heat in order to work properly. Remember, heat kills those pathogenic organisms (disease-causing bacteria and viruses) and hastens the composting process.

If the heap is built too high or has too much soil, the underlying layers will be compressed, aeration will be poor, and the process slowed down or stopped. If the heap is not tall enough, heat and moisture loss will be so great that results are likely to be poor.

If the width and length of the heap at the bottom are too short, heat and moisture loss will also occur. Since the outer 6-inch layer dries out and cools off, it does not undergo composting; however, it does serve as an insulating layer.

The *critical size,* then, of an outdoor compost heap or pile (not in bins) is 42 to 48 inches high by 8 to 12 feet wide and 8 to 12 feet long or more. This critical size can be reduced for backyard composting to about 3 feet by 3 feet by 3 feet if a bin is used with walls that have an insulating effect. Suitable walls, in order of desirability, are 2 x 6 inch lumber, concrete blocks, brick, or woven picket fencing (or welder wire) lined on the inside with a ¼- to ½-inch thickness of cardboard or old newspaper.

In warm, humid climates, if the bin is filled from May to July, natural humidity, rainfall, and warmth may make it unnecessary to line a wire bin. But in the fall, insulation will be important anywhere; otherwise the composting process will slow down or stop until the outdoor temperatures of spring get them going again.

In more western areas, insulating and watering are important during any season, for humidity is often so low a small heap in a wire bin without a lining dries out almost completely.

Carbon-to-Nitrogen Ratio

If you build a compost heap and it fails to heat up, it could be lack of moisture or too much moisture, lack of aeration, or it just might be the wrong carbon-to-nitrogen ratio. Microbes won't work unless you feed them right. To a microbe, proper feeding means a C:N ratio of 30:1; up to 35:1 is fairly good. Maybe you built your heap with a 50:1 ratio, which is too lean in nitrogen.

When you build a heap, use materials with narrow C:N ratios; *and* wide C:N ratios to get a balanced mix. For example, if you are going to use raw sawdust (which is 511:1), you'd better use lots of something like young grass clippings or sewage sludge to bring that ratio down (Table 10).

Table 10
Carbon-to-Nitrogen Ratios of Some Organic Materials*

	Nitrogen (%)	Carbon:Nitrogen Ratio
Activated sewage sludge	5.0-6.0	6:1
Grass clippings (fresh)	4.0	12:1
Grass clippings (dried)	2.4	19:1
Manure, farmyard	2.2	14:1
Straw (oat)	1.1	48:1
Straw (wheat)	0.3	128:1
Sawdust (aged)	0.25	208:1
Sawdust (fresh)	0.11	511:1

*From H.B. Gotaas, *Composting,* American Public Health Association, New York, N.Y., 1956.

Ways To Make Your Compost

There is a "fast method" that takes 14 days and considerable labor along with grinding and mixing of materials. And there is a slow method that requires 6 to 9 months but takes little or no follow-up labor and no grinding or mixing.

The Fast Method (14 Days)

1. Assemble plant residues, coffee grounds, vacuum cleaner sweepings, manure, hair clippings, waste paper and other decomposable materials. Keep out mushy garbage, tin cans, rubber and plastic, glass, woody prunings, salt and chemicals.
2. Use a compost grinder to grind above materials to dimensions 1½ inches each way or finer. As you grind, mix a variety of materials. Be sure to mix in some of the higher-nitrogen materials such as manure.
3. Hose down the material as you grind it to raise the water content to about 50 percent (2 lbs. of moist material will weigh about 1 lb. when the water is removed).
4. On the bottom of the compost bin place a layer of coarse, unground material, such as old okra or corn stalks, 4 to 6 inches deep. This allows good air circulation from the bottom.
5. Use a flat-point shovel or bucket to place mixed, ground material in bin. If manure is available, mix 1 part of manure to 3 parts (by volume) of plant material.
6. In low rainfall areas, make the top of the compost heap in the bin flat to catch rainfall or sprinkler irrigation. In high rainfall areas it may be desirable to slant the top like a roof to shed excess water and prevent the heap from becoming soggy.
7. Cover completed heap with burlap bags and hold corners down with bricks. Or, cover with a 6-inch layer of leaves, straw, etc., to hold moisture in and insulate against heat loss. In windy areas it may be necessary to use erosion netting (or a bird net used to keep birds out of trees bearing fruit) anchored at all four corners to hold the top insulating cover in place.
8. Turning: Turn every 4 days with pitchfork or other tool to lift compost material out of bin and drop in an adjoining bin. Place outside layers of heap on the inside of the new heap. Turn on the fourth, eighth, and twelfth days. This ought to do it. (You won't have to jog for exercise on those particular days!)
9. Shrinkage of heap: Watch for heap to shrink in volume to about 50 percent. This should signal the completion of the composting process and occur in about 2 weeks.
10. Reminders: This method won't work unless materials are *ground* or are naturally in small pieces. Old tomato vines, corn stalks, and the like cannot be turned so must be used with the "slow method" below.

The "Slow, No-Turn Method" (6 to 9 Months)

1. Assemble plant residues, manure, and other refuse.
2. Place a coarse bottom layer of old stalks about 4 inches deep on the bottom and then a 2-inch layer of finer material.
3. Spread a 2-inch layer of manure.
4. Sprinkle on a little compost or rich soil.
5. Wet down with hose.
6. Add a 6-inch layer of plant residues, then repeat steps 3 through 6 until the heap is about 34 inches high.
7. Cap off the top with leaves, grass clippings or other fine-textured residues and anchor down if in windy country.
8. Wait. No turning is done unless you want to speed up the process. If viny residues such as tomato vines or corn stalks are present, it will be very difficult or impossible to turn anyway. After 2 to 3 months, when partial decomposition has broken down some of the residues, it may be possible to turn the heap. Be sure to leave an empty bin adjacent to pitch the compost into.

The Wire Compost Bin

A very inexpensive, easy-to-construct compost bin can be made of steel wire to form circular bins 3 feet high and 3 to 4 feet in diameter. (A 3-foot bin requires a piece of wire 3 feet high by about 10 feet long.) A smaller bin (3 feet diameter) surrounded by a larger one (4 feet diameter) works best because you can fill the gap between bins with insulating material. A 2-inch mesh or small mesh is preferable to keep materials from falling out. Procedure for construction is as follows:

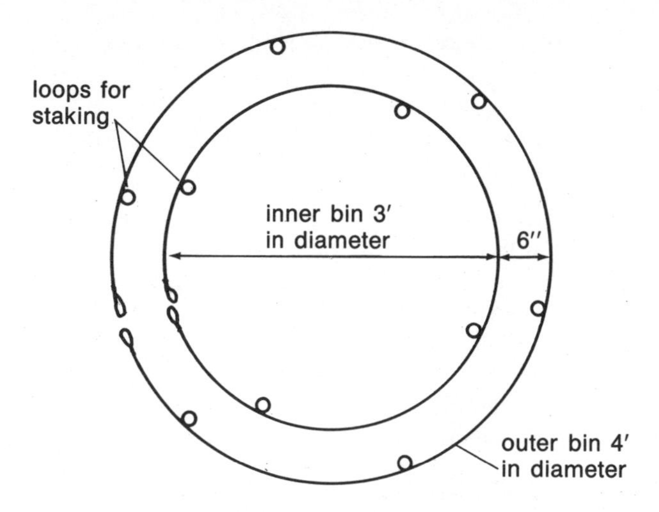

Figure 14. *Top view of wire compost bins with ends ready to pin together.*

1. Use small bolt cutters to cut off a piece of wire 10 feet long for a 3-foot diameter bin or 12 feet for a 4-foot diameter bin (a 4-foot diameter may be better). Make the cuts halfway between the horizontal wires to leave prongs for bending into loops.
2. Form the piece of wire into a circle about 4 feet in diameter with the prongs facing each other.
3. With heavy pliers, grasp one of the prongs at the top and form it into a loop. Then twist the loop to make it lie in a horizontal plane. Do the opposite prong the same way.
4. Repeat step three every 8 or 10 inches and end at the bottom set of prongs.
5. Fasten the ends of the wire prongs together using the loops by (a) inserting a wood dowel or steel rod about 38 inches long, or (b) tying prongs together temporarily with short pieces of wire.
6. Fill the bin with compost materials.
7. For turning or harvesting, pull the rod upward and out, and spread the ends of the wire bin apart for access to the heap. (Figure 14). For turning, have an empty bin set up nearby.
8. Line inside of bin. It may be helpful to line the inside of the wire bin with old newspaper, old wrapping paper or cardboard to conserve heat yet admit air. Heat loss at the edges of a compost heap may stop the composting process there; drying out will do the same thing.

Picket Fencing for Compost Bins

Woven wood picket fencing comes in 50-foot rolls 3 feet high. Decide what diameter bin you want and multiply this figure by 3. The result will be the length of piece to cut off to make the bin. Use the same method of the loops and dowel to close and open the two ends of the bin.

Double-Wall Wire Bin

A single wall bin of wire will never have good insulating properties. To add this feature, make two bins, one of them a foot smaller in diameter to fit inside the larger bin (Figure 14).

To load, set up the small bin and fill it. Then install the outer bin and fill the 6-inch space between the walls with balls of old newspaper or straw.

New Zealand Compost Bin

This bin is simply a square bin 39 inches high by 4½ feet square made of wooden boards. Two sides and the back are held together by long bolts with wing nuts. The front is of removable boards. The two sides at the front are held in place by an "anti-spread" board (a piece of 1-inch by 2-inch board 5 feet long with notches cut at 4½ feet apart near each end. When laid across the top at the front, the sides of the bin fit into the notches and are held fast.

Harvesting Finished Compost

No one "makes" compost. What you make is a compost *heap.* This heap can be inside or outside a bin. While it is undergoing the process of decomposition (and heating up, we hope), it is composting. When this process is finished and you cannot tell one part of the material in the bin from another part, it is finished compost, or just "compost." Nearly any compost heap at harvest time will have some undecomposed material.

Compost is harvested like any other crop. Here are the steps:

1. Assemble your harvesting equipment. A heavy-duty wheelbarrow is handy, though not essential. A shovel is a necessity, as is a screen or sieve. You also need a place to store the harvested material—plastic bags or plastic garbage cans. A piece of plywood about 4 feet square is handy for the screened compost to fall on. Otherwise, use a tarp.

Compost screen or sieve 2' wide by 4-6' long built of 1" × 4" lumber. One-inch mesh poultry netting is attached to the bottom with screen molding.

2. Lean the sieve at a 45-degree angle against some support convenient for shoveling from the bin to the sieve.
3. Rake aside the top layer of material on the compost heap and shovel out the compost 1 shovelful at a time, dropping it on top of the sieve. Let gravity roll it downward.
4. When several gallons of screened compost have collected on the board below, use a flat-point shovel to shovel it into cans or sacks.
5. Place the material that would not pass through the screen in a separate pile to be recycled back into the next heap.
6. Store the screened compost where it will be protected from sun and wind. Or, place it on a piece of plastic under a tree and cover with another piece of plastic using bricks to hold down the corners.
7. Treat compost for white grub infestation. The adult May beetle or the June beetle love to lay their eggs in compost piles, so when you spread the compost in your garden, you may also seed it with grub worms. You can either treat it for grubs or prevent infestation as follows:

—Make wire screen covers for your compost bins to exclude the egg-laying beetles, or
—Place the screened compost in a plastic garbage can with a tightly fitting cover. Place an old jar lid on top of the compost, and in the lid place ¼ cup of moth balls. Put the cover on and place can in the sun for a week. Be sure top is airtight. To ensure this, cover the open can with a layer or two of plastic sheeting, then push the top on. The fit should then be snug. After a week or two, remove the top for airing.

Note: If piles of organic material were not lying about during the egg-laying season (May or June) and you harvest compost before the end of April, there should be no grubs or eggs in your compost. The same is true if you filled an empty bin with fresh organic matter that had not been lying around in piles from July on.

Building Your Own Concrete Block Bins

This type of bin is expensive but will be trouble-free and will last a lifetime. An ideal set-up is to have a set of three bins each 4 feet square, side-by-side. Dig trenches, place in a layer of sand 2 to 4 inches thick, compact and level the sand. Arrange it so the first course of blocks are buried halfway under the ground line. Make the walls six blocks high. Use whole and half blocks to be able to "break the joints." Make the inside dimensions about 48 inches each way.

The last step is fixing up the fronts of the three bins so boards can be slipped in and out for easy loading or unloading. The best approach is to build a front groove assembly out of 2 x 2s and 1 x 4s nailed together. These need to be long enough to be put into the ground 12 to 18 inches and have concrete poured in the hole around each post while it is held perfectly straight up and down. These posts with groove assembly will be free-standing and touching but not attached to the bin (Figure 15). Rough, unplaned boards an inch or more thick by 6 to 12 inches wide may be used for the front boards (2 x 6s or 2 x 8s work well).

Using Compost

Since compost is usually in short supply, the best use is usually spot treatment. Use it to cover

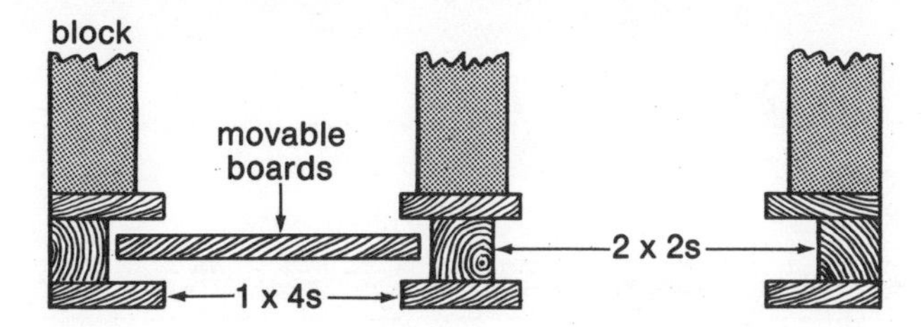

Figure 15. Top view of front groove assembly for concrete block compost bins.

The best way to use compost when supplies are limited—these two adjacent rows of sweet corn seed just planted are covered with an inch of compost.

seeds in the planting trench or to fill in around transplants.

Use of Compost for Disease Control

Nematodes are a serious problem in many areas, as are various soil-borne organisms that are associated with root rot. The following is from the November 1975 issue of *Organic Gardening* magazine:

" 'Greenhouse and nursery growers can reduce, and in some crops eliminate, the use of soil fungicides and hazardous fumigants.' That's the conclusion of plant pathologists Harry Hoitink and associates at the Ohio Research and Development Center in Wooster, Ohio. The reason? Composted hardwood bark used instead of peat in potting soil mixtures controls root rots and other soil-borne diseases. 'Phytophthora, phythium and rhizotonia root rots are important diseases of several woody ornamentals produced in various peat container media,' says Dr. Hoitink. These root diseases do not occur on similar plants produced in composted hardwood bark media.

"To test the hypothesis that composting eradicates the pathogens, Hoitink buried infected roots in a pile of bark before composting in a bark pile that was not composted. After 10 weeks, no pathogens could be recovered from the composted bark, but they survived at high levels in the non-composted bark pile. 'And the composted bark inhibits the recolonization of the disease fungi too,' said Hoitink.

"Would composted bark work the same way for field-grown plants? Hoitink thinks so. 'Composted bark incorporated into phytophtora-infested soils in Oregon at rates of 40 to 100 tons per acre provided a high degree of control of red stele disease of strawberries,' he says. 'And we know that certain diseases like root knot on tomatoes caused by nematodes are also reduced when bark compost is applied.'

"Will other compost work? Hoitink thinks not, but he does think the secret is in exact and meticulous composting as much as in the material used. A pile of manure that has slowly rotted over the years won't do it. His description of the correct way to compost is the same as the methods described in this book. 'Apparently, the heat kills the pathogens, but we don't know that for sure,' he said.

" 'At least half the nurserymen in Ohio are trying the bark and are enthusiastic about it,' Hoitink added. 'Not only can it mean reduced costs from less use of soil fungicides and fumigants. We can also utilize the huge piles of bark now a waste byproduct of the paper pulp industry. We can stop using so much peat and thereby reduce landscape destruction of peat bogs.' "

Nematodes have become an even more acute problem now that the EPA has taken "Nemagone" and "Fumasome" off the market. These two materials were used to control nematodes around the roots of living plants. No other nematode-killers available will do this. Perhaps compost is a solution.

Common Questions About Composting

1. Is extra nitrogen necessary in making a compost heap?

Not necessarily, though it may take a little longer with low-nitrogen material. Since good compost has been made with vegetable trimmings and 21 percent paper and a C:N ratio of 55:1, it would seem that no extra nitrogen would be needed in most cases. If you have met all other requirements and your heap refuses to break down, make a shallow basin on top of your heap, broadcast a light sprinkling of water-soluble nitrogen fertilizer, and water it in.

2. You said microorganisms in the compost heap need a diet of 30-35 parts of carbon to 1 part of nitrogen but also that good compost had been made with 55 parts of carbon to 1 of nitrogen. How is this possible?

Probably by free-living nitrogen-fixing microorganisms. Two of these are *Azotobacter* and *Clostridium butricum.*

The air around us is about 80 percent nitrogen. The nitrogen-fixing organisms can utilize (or fix) this gaseous nitrogen and build it into their tissue under proper conditions. These conditions include (1) low available nitrogen, (2) presence of organic matter (carbon), (3) a pH of 6.5-7.5, and (4) good aeration (to furnish the oxygen and nitrogen for fixation). This indicates that you may be able to "starve" a compost heap for nitrogen to make the nitrogen fixers go to work. No experimental evidence exists, however, to show compost made successfully from materials with a C:N ratio higher than 55:1, so caution should be observed in using too much woody material such as all sawdust or all wheat straw (C:N ratios of 128 to 208).

3. Is soil necessary?

No. A comprehensive California study* showed that municipal refuse shredded without any soil added made good compost.

4. Are bins necessary?

No, but they are highly desirable for (1) tidiness, (2) keeping flies away in some cases, and (3) to insulate the outer layers form heat and water loss. Bins 4 feet square by 3 feet high made of concrete blocks with a removable board front in sets of three are ideal. Without bins, the outer 6 inches of the heap do not convert to compost. Bins are necessary unless you build a huge heap.

5. What about activators?

Plant residues are covered naturally with the necessary organisms, so no activator is needed.

*"Reclamation of Municipal Refuse by Composting," Sanitary Engineering Dept., University of California, Berkeley, Cal., 1953.

6. How about adding limestone?

This may be necessary in acid soil areas. Only when the compost heap has a pH of 5.5 or 6.0 is it necessary to add limestone. In acid soil areas, sprinkle on 1 lb. of limestone per 20 square feet.

7. How long does it take?

If ground and mixed and turned (aerated) every 3-4 days, you can have compost in 14 days; by just piling up garden residues in a bin and keeping moist with no turning, 6-9 months.

8. If I do need to add extra nitrogen, what do I add and how much?

Spread about 1 lb. of ammonium sulfate (21-0-0) plus ½ lb. of superphosphate per 20 square feet of surface when layers are 6 inches thick. This is equal to about 0.2 lb. of actual nitrogen and about 0.1 lb. of actual phosphorus.

9. Is animal manure necessary?

No, but it is a very suitable ingredient for the compost heap and it has the extra nitrogen often needed to balance out low-nitrogen materials such as straw or newspaper, plus organisms that act as activators.

10. Are ventilation holes necessary?

Not if the heap is made correctly.

11. Do flies around my compost heap mean I am furnishing a breeding ground for them?

No, not if the compost heap is properly made. Flies will rest on a compost pile but do not breed in it. Fly larvae placed in compost material do not develop where the lethal temperature of 122 degrees for fly larvae is reached overnight.

12. Can sewage sludge be safely used?

Only in non-industrial areas. Both raw and digested sewage sludge have been safely used with good results when accompanied by temperatures above the thermal death point of pathogens within 3 days. Not over 10-20 percent (by weight) of sludge (solids and/or liquids) should be used.

13. How hot should a compost heap get?

The third day it should heat up to about 158 degrees. This is equal to a cup of hot freshly poured coffee which you must sip slowly to avoid burning lips and mouth. The next day or two the temperature will drop to about 110 degrees or like lukewarm coffee.

14. How do I determine whether or not to add extra nitrogen to the heap?

An average carbon-to-nitrogen ratio of 30:1 to 35:1 is necessary. The C:N ratio of various organic materials are shown in Table 10, page 33.

Inorganic and Organic Fertilizers

Inorganic Fertilizers

What is a fertilizer? According to the *Western Fertilizer Handbook,* "A fertilizer is any natural or manufactured material added to the soil to supply one or more plant nutrients."

To use fertilizers wisely, we need to learn a little chemistry and agree on some terminology.

What do we mean by "chemical" fertilizer or "commercial" fertilizer? Is there a non-chemical fertilizer or a fertilizer that cannot be sold commercially? What is "gardening with chemicals"? Are chemicals evil? It all depends on what you mean by these terms.

Chemistry is the science of the composition of matter. Matter is anything that has weight and occupies space. The three kinds of matter are gases, liquids, and solids.

Matter is made up of building blocks called chemical elements. About 100 of these elements have been discovered by science. Everything in our physical universe is made up of one or more of these elements or "chemicals."

The term "chemical" fertilizer has little meaning. Compost is as much "chemical fertilizer" as is a sack of superphosphate, since both are made up of chemicals (chemical elements).

The term "commercial fertilizer" also has little meaning. A sack of compost, a sack of leaf mold, a sack of cottonseed meal—all are organic fertilizers and all are found in the channels of commerce. They are bought and sold as freely in the marketplace as are sacks of ammonium sulfate.

Of particular interest to the gardener are not all the 100 chemical elements but the "Magic 13" that cause plants to grow. The chemist has set up a list of abbreviations for each of these elements. N, for example, stands for nitrogen, and this is logical since the word "nitrogen" begins with "n." P stands for phosphorus, also logical. But here the logic ends. K stands for potassium.

Another piece of chemical nonsense is that fertilizer bags do not show some elements as they really are. Phosphorus, for example, will be shown on the bag as phosphorus pentoxide, P_2O_5 (which is called *phosphate*). Potassium is shown on the bag not as the element potassium, but as potassium oxide, K_2O (which is called *potash*).

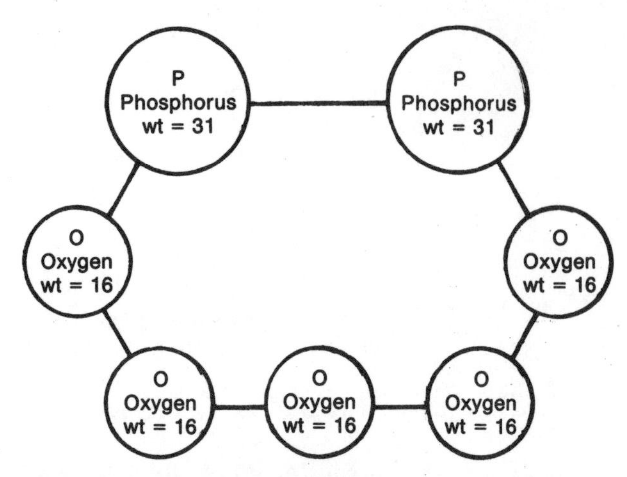

***Figure 16.** A phosphate molecule (P_2O_5). Because of its heavier atomic weight, phosphorus comprises 44 percent of the molecule, even though there are 5 parts of oxygen to 2 of phosphorus.*

A typical bag of fertilizer will have three numbers printed on the bag such as 5-10-5. These three numbers are percentages. They tell us how much of the "Big 3" are in the bag. Since percent means how much per hundred, 5-10-5 means 5 percent nitrogen, 10 percent phosphate, and 5 percent potash.

If there are only 15 percent active ingredients, as in a bag of 5-10-5 ($5 + 10 + 5 = 15$), then what is the other 85 percent? It is inert or "filler" material made up of other chemical elements such as carbon, hydrogen, oxygen, and water (water is hydrogen and oxygen).

A typical bag of fertilizer weighs 40 lbs. Five percent will be nitrogen. The 40-lb. bag with 5 percent nitrogen will contain 2 lbs. of actual nitrogen.

Phosphate vs. Phosphorus

Phosphorus (P) is one of the "Big 3" essential plant food elements, but as described on pages 25-26, it just won't put up with being alone and at the first opportunity combines with oxygen to form an oxide. The unit formed has 5 particles of oxygen hooked up to 2 particles of phosphorus, so we call it a "pentoxide" from the word "penta," meaning five (Figure 16). This is phosphate.

The second number on the fertilizer bag shows the percent of phosphate present, such as the "10" in a bag of 5-10-5. In a 40-lb. bag of 5-10-5, 10 percent of the weight will be phosphate or 4 lbs. (10 percent of $40 = 4$).

But this 4 lbs. is not all phosphorus. In fact, less than half of it (44 percent) is actual phosphorus.

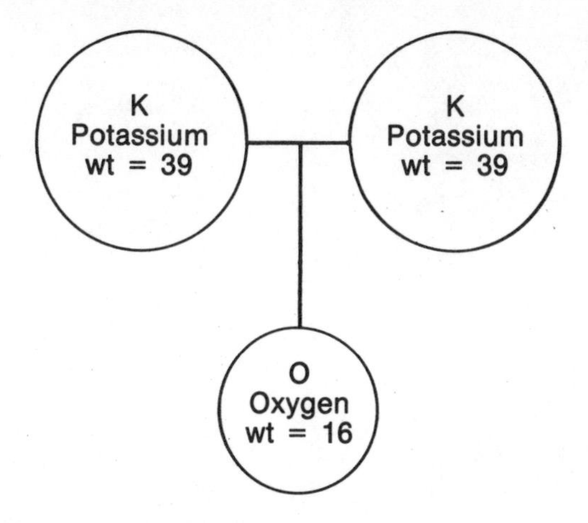

Figure 17. *Molecule of potash (K_2O). Potassium comprises 83 percent of this molecule.*

Why? Because the P_2O_5 unit has 5 particles of oxygen tagging along. So in the end, the 40-lb. bag of fertilizer contains only about 1.8 lbs. of actual phosphorus (44 percent of 4 = 1.8).*

Potash vs. Potassium

Potassium is abbreviated "K." It, too, is one of the "Big 3" elements. It also loves company and comes combined with oxygen (Figure 17). The fertilizer bag usually expresses the potassium as K_2O, or potash.

In a 40-lb. bag of 5-10-5, 5 percent, or 2 lbs., will be potash. But potash contains both potassium and oxygen, so only 83 percent of this unit is actual potassium (Figure 17)—the 40-lb. bag contains only about 1.7 lbs. of actual potassium (83 percent of 2 = 1.7).

Chelated Fertilizers

If you think phosphate is complicated, consider the new chelated fertilizers. Iron chelate is a typical example. It has a particle of iron in the center of a group of other elements shielding it from soil reactions to the outside (Figure 18). The salt-type form of iron, iron sulfate, has an identical particle of iron but no shielding (Figure 18B).

The iron in iron fertilizer is like a quarterback on the football field. If he has several good backs to block for him and shield him from the opposing tacklers, he may run a long way. But if the ball carrier has no blockers to shield him, he won't get far (here the quarterback is like the iron in iron sulfate).

*This percentage is based on respective atomic weights of phosphorus and oxygen. Phosphorus has an atomic weight of 30.9; oxygen, 16.

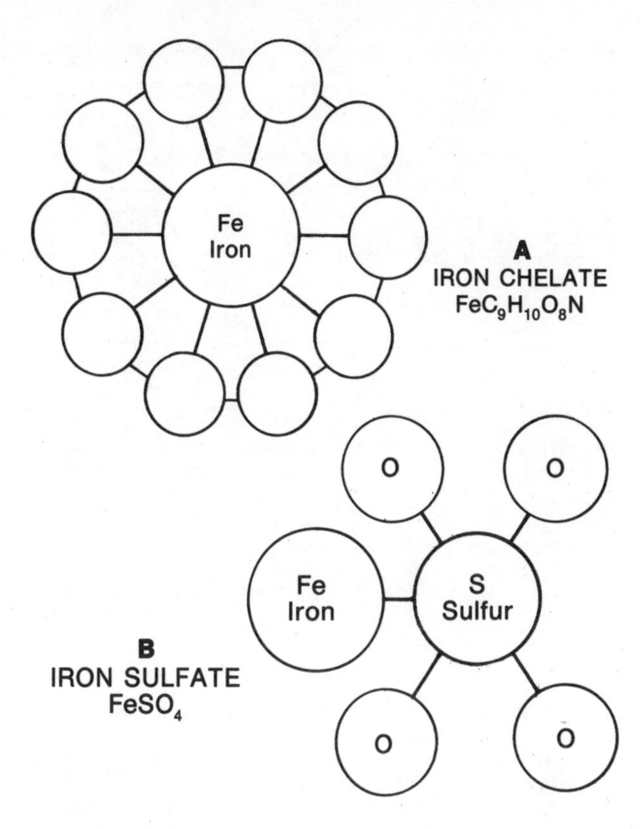

Figure 18. *Organic chelated iron fertilizer (A) vs. inorganic salt-type iron fertilizer (B). The iron particle in A is protected by a ring of other elements (carbon, oxygen, hydrogen, and nitrogen).*

Nails for Iron-Hungry Plants?

Since nails are made from iron and plants can suffer from iron deficiency, it seems logical to throw some old rusty nails around the roots to solve the problem. Sorry, but it won't work. You have added elemental iron, Fe, to the soil. The first thing you know, the iron rusts and forms iron oxide. Iron oxide will not dissolve in water, so the plant roots cannot get it.

Plant Food Elements and What They Do

By watching how your plants grow, you can often tell if they're being fed properly. A tomato plant, for example, that is big and beautiful but bears no fruit is about as useful as a hammer without a handle. It may not bear because of weather conditions, or it may be too much nitrogen. Excessive nitrogen promotes luxuriant growth of stems and leaves but may discourage

A centrifugal broadcaster is a good way to apply fertilizer over large areas. Powders work poorly in this kind of spreader; granules or pellets work much better.

fruit production. If you are growing cabbage or lettuce or some other crop whose stems and leaves are edible, this would be no problem. Too little nitrogen, on the other hand, may cause leaf yellowing and stunted plants.

If you want a root stimulator and quick, early growth, flower or fruits, look to phosphorus. Starter solutions and "root stimulator" solutions are simply a mixture of a high-phosphorus fertilizer and water. You can make your own by mixing 2 tablespoons of superphosphate per gallon of water.

Phosphorus also helps grow quality crops and promotes disease resistance.

What Analysis Fertilizer Should You Buy?

Go into any garden center and you'll see an amazing array of fertilizers for sale. There will be a "Rose Food," "African Violet Food," "Pecan Food," "Houseplant Food" and so on. Each of these bags will show (though sometimes you have to look hard) a different analysis such as 10-5-5, 5-10-5, 16-8-8, 21-0-0, 0-20-0, etc.

Is there a special analysis that makes a superior fertilizer for roses so that we can call it a "rose food"? Do certain plants need certain fertilizer mixes? I doubt it. What we are dealing with here are fertilizer ratios—the relative amounts of nitrogen, phosphate, and potash in a bag of fertilizer. A bag of 5-10-5, for example, has a ratio of 1:2:1, or 1 part of nitrogen to 2 parts of phosphate to 1 part of potash.

The trouble with labeling any kind of fertilizer as a food for certain plants is that there are too

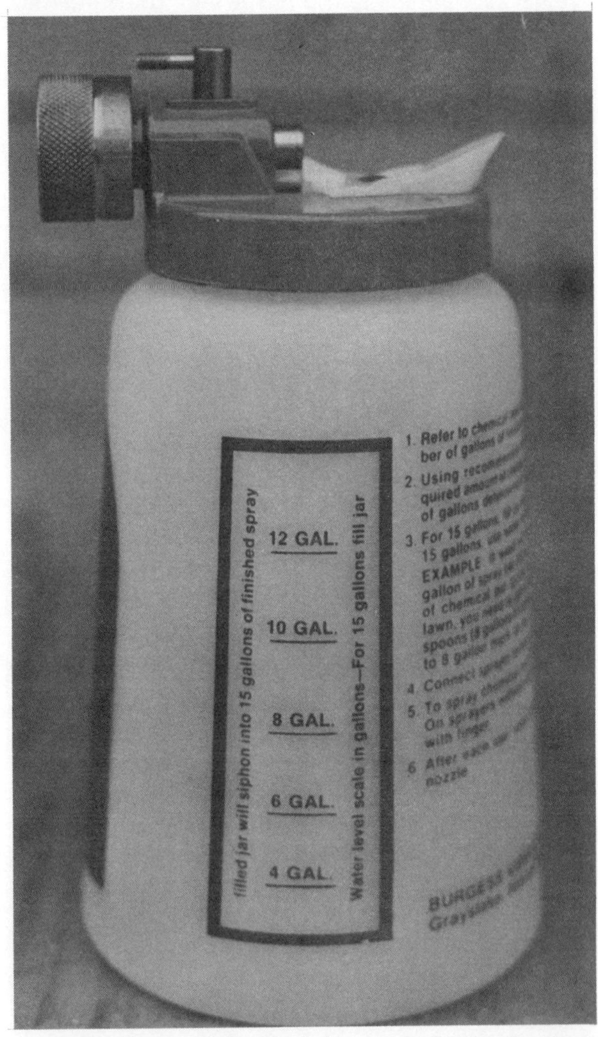

A garden hose applicator is the perfect thing for liquid fertilizers. It attaches to the end of the hose, and applies fertilizer as you water.

many other factors involved—especially the soil and what it already has in it.

Plant Composition and Fertilizer Ratios

Why not analyze the plant—see what it contains, and how much—and then tailor a fertilizer to fit these ratios? This is a logical approach. It's been done, and generally the ratios tend to be 2:1:2 or 3:1:3, showing high nitrogen, high potassium and low phosphorus (Table 12, page 46).

The trouble is, it's one thing to add a tailor-made fertilizer to the soil but quite another to get the plant to take it up as you applied it. In this case you would be applying a relatively small amount of phosphorus to start with, and by the time soil reactions finish working over the phosphorus you applied, very little would be taken up. The only way to beat this is to add chelated phosphorus or phosphorus as organic fertilizer. Because of the vulnerability of phosphorus when applied to the soil, the fertilizer analysis to use when phosphorus is needed is a 1:1:1 or 1:2:1 (12-12-12 or 5-10-5, for example).

A timed-release fertilizer is probably the most handy type for container plants. It comes in small pellets that can be broadcast by hand or used in a broadcaster for covering large areas.

Soil and Crop Testing for Fertilizer Ratios

The ideal way to get out of the confusion over which kind of fertilizer to buy is to test the soil and fertilize based on the results. Then check plant response to see if it worked. But this is research and takes lots of time and money. And not much of it has been done.

So what is the home gardener to do? You can send off a soil sample for chemical analysis. Most state agricultural universities perform this service at a nominal cost. *Call your county Extension agent for information.*

How to take a soil sample. You need a trowel, a plastic bucket or bowl, and a small plastic bag.

The best sample is a *composite* sample—one made up of a mixture of several other samples. In a garden 30 feet by 30 feet, walk across it diagonally and take about four samples more or less evenly spaced.

To take an individual sample, use the trowel to rake off the top ¼ inch, then dig out a trowel-full and cast it aside. Now take the actual sample. Cut a slice of soil off the side of the hole you dug and place it in the bucket. Go down to about 6 inches deep. Repeat until you have worked your way across the garden area (Figure 19).

Now break up all the slices of soil and mix them. Take about a pint of the mixture and place in a plastic bag. If you take more than one sample, be sure to identify each sample by a note inside the bag.

The last step is to package the soil sample for shipping. A small cardboard box will do. Most county Extension offices have boxes for this purpose.

Interpreting soil test results. When your soil test results come back, they will show the soil pH and chemical analysis of essential elements. Usually, the nitrogen, phosphorus, and potassium will be shown in terms like "low" or "high" and a general fertilizer recommendation will be made.

The idea of the soil test is to inventory what your soil already has, then add only what you need. This means you can't really say a given fertilizer is a "Rose Food." It depends on the soil where the roses are to be grown. The only exception would be with soilless culture, where there's nothing except what you put there. For this purpose, there is a universal liquid fertilizer—called the Steiner solution—that is excellent.

Without soil test results about all you can do to fit the fertilizer to the plant is use a well-balanced fertilizer. Probably the best choice is an even ratio

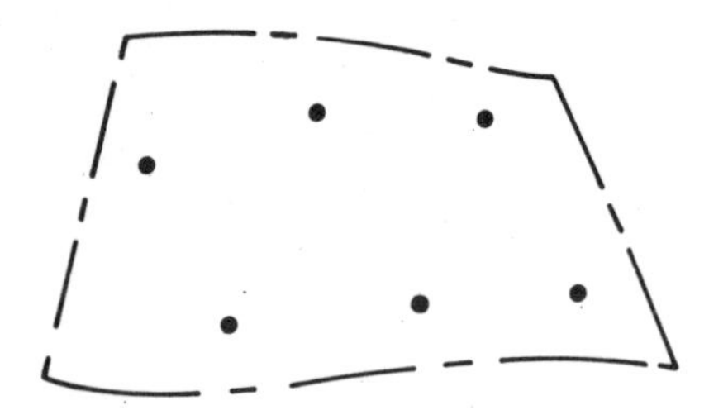

Step 1.
For a composite sample, obtain at least 4 samples from the area, put in a clean container, and mix thoroughly. Take out about 1 pint. Place in a clean container such as an ice cream carton or paper sack, and submit for testing.

Step 2.
To take a sample, use a garden trowel or spade. Dig a V-shaped hole, then take a ½-inch slice from the smooth side of the hole. Place in a bucket and repeat in about ten places. Take samples to a depth of 6 inches in flower beds and gardens and 4 inches in lawns. A soil probe or soil auger also can be used.

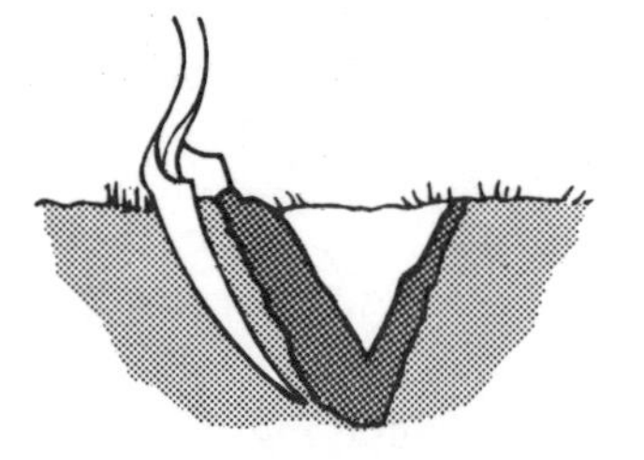

Step 3.
Label each sample with number and name. Be sure to keep a record of the area from which the samples came.

Figure 19. *How to take a soil sample.*

such as 8-8-8 or 10-10-10. The three numbers do not have to be exact, only approximate—for example, a 10-8-8 is close enough to an 8-8-8.

If you're growing a plant to produce lots of flowers, a fertilizer a little higher in phosphate (a 1:2:1 ratio) might be better—a 5-10-5 is good.

One way to find out which fertilizer is best is to run some simple tests. Fertilize a few plants with one fertilizer and a few more plants with another and see which gives the best response. If there's no difference, one fertilizer will be as good as the other. But be sure you run your test on plants of the same variety and size.

Another fairly safe way is to forget the labels that say "Tomato Food" and the like and go to organic fertilizers. These have all 13 of the essential elements, and generally, they are in good balance for plant growth. (See pages 47-49).

Fertilizers to Have On Hand

Most home gardeners will find it handy to have the following in 40 to 50 lb. sacks:

1. *Ammonium sulfate (21-0-0):* This is a nitrogen-only fertilizer, a salt-type fertilizer, and is handy for lawn fertilization and other situations where all you need is nitrogen.
2. *Superphosphate (0-20-0):* A phosphorus-only fertilizer handy for making starter solutions and stimulating flower, fruit and seed formation.
3. *Ammonium phosphate (16-20-0):* A nitrogen-and-phosphorus combination for use where these two elements are needed.
4. *Cottonseed meal:* A balanced, organic, slow-release fertilizer with an analysis of 6.5-3.0-1.5, plus all the trace elements. (See Table 13, page 47.)
5. *Ferrous sulfate:* Use only in alkaline soil areas. This is iron sulfate, sometimes called copperas. It contains 20 percent iron in a soluble form available to plants if not tied up by soil reactions.
6. *Dolomite:* This is limestone that also contains magnesium. Helpful in mixing with peatmoss to raise the pH when making up potting soils such as peatlite.
7. *Steiner solution:* This is an 8-5-16 analysis with all the trace elements in proper proportion for feeding patio plants in containers, houseplants, or home greenhouse sand culture. One tradename for this is "Pronto Hydroponic Special," available from Pronto Fertilizer Co., P.O. Drawer 247, Wisner, La. 71219.

The Salt Index

You can mix superphosphate in a planting trench and the seed will come up normally. But mix ammonium sulfate in the trench with the seed and they'll die. Why? Because the salt indexes of nitrogen fertilizers (such as ammonium sulfate) and potassium fertilizers are very high. These fertilizers have salt indexes nearly eight times greater than superphosphate (Table 11).

What is this "salt index"? It's a number that tells us how much a given fertilizer will increase the saltiness of the soil solution. Since roots grow in this solution, they can shrivel up and die, killing the plant, if it gets too salty due to application of salt-type fertilizers.

The most dangerous time for injury from fertilizers with high salt indexes is during seed germination. This is why we sidedress with a nitrogen fertilizer but do not apply it to the planting spot at planting time.

The more natural-type fertilizers tend to have low salt indexes—calcium fertilizers like limestone and gypsum are examples (Table 11).

The salt index gives us a clue as to placement of salt-type fertilizers. If the salt index is high, keep the fertilizer away from germinating seeds.

Gypsum

Gypsum is often mentioned as a good amendment to improve soils. Chemically, gypsum is calcium sulfate or $CaSO_4 \cdot (2\ H_2O)$. It is a calcium and sulfur fertilizer.

Gypsum is a white or yellowish-white mineral that occurs in beds. It is mined, ground and put in sacks. It is also called plaster of Paris.

Gypsum may be used as a soil conditioner, especially in clay-type soils, which are often high in sodium. The sodium makes for a tight soil that won't take water. Gypsum worked into these soils can be very beneficial—it helps convert the insoluble sodium into a soluble form that can be leached out of the soil. Good-quality irrigation water or rainfall will dissolve the sodium and leach it down and out of the root zone. Gypsum soil treatment works only if you follow up with leaching.

The Weed-and-Feed Madness

As a professional horticulturist I have frequently been called out to look at a sick tree with leaves all curled and twisted. "Did you apply a herbicide around here lately?" I ask the owner.

"No, just fertilizer," he says.

What did it say on the bag?" I ask.

Table 11
Composition of Home Gardening Inorganic Fertilizers and Their Salt Indexes

Fertilizer	Analysis*	Salt Index (per 20 lbs. of plant food elements)
Nitrogen-rich types		
Ammonium nitrate	35-0-0	3.0
Ammonium sulfate	21-0-0	3.2
Ammonium phosphate	16-20-0	1.6
Ammonium phosphate	11-48-0	2.4
Potassium nitrate	14-0-46	1.6
Sodium nitrate	16-0-0	6.0
Urea	46-0-0	1.6
Phosphorus-rich types		
"Single" superphosphate	0-20-0	0.4
"Treble" superphosphate	0-46-0	0.2
Ammonium phosphate	16-20-0	1.6
Ammonium phosphate	11-48-0	2.4
Potassium-rich types		
Potassium chloride	0-0-60	1.9
Potassium nitrate	14-0-46	1.6
Potassium sulfate	0-0-54	0.8
Potassium magnesium sulfate	0-0-22	2.0
Miscellaneous materials		
Limestone		0.08
Gypsum		.25
Table Salt		2.9

*Percentages of nitrogen, phosphate and potash, respectively.

He doesn't remember, and when we finally find the sack, my suspicions are confirmed: he has applied a weed-and-feed type of fertilizer. This type of fertilizer is a mixture of plant food elements and one or more herbicides (weed-killers).

What's wrong with this? Plenty. It's like playing Russian roulette with your plants. With few exceptions, homeowners have no business applying weed-killers around home grounds. Why? These powerful chemicals are just too dangerous to plant life. Like a double-edged sword, they easily cut the user as well as the enemy.

Herbicides are *plant* killers. Highly concentrated, a little goes a long way. Very careful dosage is necessary—some soils take more than other soils; some herbicides can be used around some plants but not around others. It's a jungle situation. Unless you're an expert, don't mess around in this jungle.

If you must use an herbicide, don't mix it with fertilizer. Apply these two separately, each to do a single, specific job. Feed your plants but avoid trying to kill other nearby plants (weeds) at the same time.

Shotgun-Mix Fertilizer Nonsense

What you are looking for in salt-type fertilizers are the "Big 3" elements. We know what these are and have the ratios figured out pretty well. Now it *seems* like a good idea to mix in a little insect-killer and a little iron and zinc for insect control and trace element feeding.

Don't do it! Avoid these "shotgun-mix" fertilizers. You can't regulate the rates of all these properly. If you base your rate on nitrogen, you may get too little iron or insecticide. If you base your rate of application on iron, you'll most likely get far too much nitrogen and kill your plants.

Let's say the fertilizer bag says 10 percent nitrogen with 0.5 percent added iron and you want to apply a pound of actual nitrogen to your lawn. Well, 10 lbs. of this fertilizer will add 1 lb. of nitrogen but only ½ lb. of iron. So what? The usual iron application is 5 lbs. per 1,000 square feet, so you have applied only one-tenth the amount needed.

Why not base the rate on how much iron is needed? If you do, go shopping for a new lawn. You'd have to apply a 100-lb. bag of the shotgun mix to add only 5 lbs. of iron. But you would also add 10 lbs. of actual nitrogen, and your lawn would be dead in a week from severe salt burn.

Don't buy shotgun mixes. Read the label and look only for N-P-K in the bag (sulfur is OK also). For iron fertilization, buy a 50-lb. bag of ferrous sulfate and apply it separately—and only when needed. For zinc problems, buy zinc sulfate or zinc chelate and apply separately.

Remember, the shotgun-mix treatment is likely to get friends along with enemies.

Fertilizer-and-Pesticide Mixes

These are another type of "shotgun mix." Sevin is sometimes mixed in with lawn fertilizers. What's wrong with this is you inoculate the bugs.

If you want to handle rattlesnakes, you can build up immunity to the venom by taking small doses over a period of time and gradually build up resistance to its toxic effects. Vaccination for disease works on the same principle.

Would you like to immunize your insect pests so that they can escape your efforts to control them? This is what you do when a little pesticide is added to the soil with fertilizer. You apply just enough to make them resistant. Then, when they build up and get really bad, you've got no knock-out punch left.

Avoid not only the "feed-and-weed" but also the "feed-and-knock-'em-dead" fertilizer mixes.

Determining Kinds and Amounts of Fertilizer Needed

How do we figure out what kind of fertilizer and how much to apply? It depends on several things. One is how much natural fertilizer is already in the soil. Soil testing is used to inventory this. Another factor is how much plant food is required to grow the crop. We call this "nutrient removal."

If you do not catch your lawn clippings but let them fall back to the soil, your lawn may fertilize itself. But usually the clippings are removed, so feeding may be necessary. If the clippings are composted and returned to the lawn, the cycle may be completed without feeding.

In a vegetable garden, every time you harvest and eat vegetables you are removing plant foods and you will probably need to add some type of fertilizer to keep your soil fertile. If you save all garden residues and give them all back to your soil, you should almost be able to make ends meet and so have to add little fertilizer.

How much plant food does it take to grow 100 square feet of vegetables? We're talking about actual pounds of plant food elements, not the entire fertilizer (we mean nitrogen, phosphate, and potash). To grow 100 square feet of snap beans, it takes about 4/10 lb. of nitrogen, 1/10 lb. of phosphate, and about 4/10 lb. of potash (Table 12).

So how much fertilizer does this require? Well, it depends on what kind of fertilizer it is. If its 5-10-5, it would take 8 lbs. (8 x .05 = 0.4). From the 10 lbs. you'd also get 1 lb. of phosphate, or 10 times as much as you need, assuming it all got taken up by the plant; and 0.4 lbs. of potash.

Table 12
Plant Food Elements Required to Grow 100 Square Feet of Several Vegetable Crops

Vegetable	Elements Needed (lbs.)			Ratio Needed by Plant
	Nitrogen	Phosphorus*	Potassium†	
Beans, snap	.4	.1	.45	4-1-4
Broccoli	.2	.07	.17	2.4-1-3.3
Cabbage	.5	.15	.6	3.3-1-4
Potato (white)	.6	.25	.8	4-1-5.3
Tomato	.6	.2	1.1	3-1-5.5

*Phosphate (P_2O_5)
†Potash (K_2O)

Look at the ratios needed by the plant. All show more or less equal amounts of nitrogen and potash but a lesser amount of phosphate. Yet most advisors say use a 1:1:1 or 1:2:1 ratio fertilizer.

Why so much phosphorus when the plant doesn't use that much? Because before the phosphorus gets inside the plant it must go into the soil, and there lots of it gets tied up. Thus, it is generally recommended to apply extra amounts so the plant will get what it needs.

If your soil contains 5 to 10 percent organic matter, this waste of phosphate may not happen. But then, if the soil is that high in organic matter, the organic matter itself will probably contain enough fertilizer to furnish all the phosphorus needed without the use of salt-type fertilizer.

Organic Fertilizers

"Organic" means "containing carbon." "Natural" organic fertilizers usually are rich in carbon, hydrogen, and oxygen and have a relatively low content of essential plant food elements.

Manure, plant residues, compost, cottonseed and similar materials are natural organic fertilizers. There are some synthetic organic fertilizers, too—urea and iron chelate, since they contain carbon, are organic fertilizers.

For our purposes, it will do to let the terms "organic fertilizer" and "organic matter" refer to materials derived from the tissue of plants and animals—the natural organic fertilizers.

If you burn 100 lbs. of dry organic matter you will end up with a small pile of ashes weighing less than 5 lbs. The other 95 lbs. went up in smoke, and it was largely carbon, hydrogen, oxygen, and nitrogen.

In the process of burning, the essential plant food elements are broken off from the carbon they were tied to. Some of the elements are left behind as ashes, but most go up in smoke. So, making your own salt-type fertilizer by burning organic matter is very wasteful since you lose not only the nitrogen but the soil-improving physical qualities of the organic matter.

In the soil organic matter "burns," but it does so in a much slower manner, which conserves nitrogen, improves soil structure, and slowly releases nutrients as well as organic acids that help make elements available that otherwise might not be.

Organic fertilizer is the ideal way to feed plants. It releases plant food slowly and steadily, the way plants need to be fed. It will not leach away. It will not burn plants.

Table 13
Composition of Organic Fertilizers*

Fertilizer	Nitrogen (%)	Phosphorus† (%)	Potassium** (%)	Ratio of Elements
Fresh Weight Basis				
Steer manure (feedlot)	1.3	1.3	1.3	1-1-1
Dairy manure	.7	.3	.6	2-1-2
Horse manure	.7	.3	.5	2-1-2
Rabbit manure	1.5	1.0	1.0	1.5-1-1
Poultry manure	1.4	1.0	.8	1.7-1.2-1
Compost	1.0	1.0	1.0	1-1-1
Dry Weight Basis				
Steer manure (feedlot)	3.5	2.3	2.7	1.5-1-1.2
Dairy manure	1.4	.6	1.2	2.1-1-2
Poultry manure	3.4	2.0	.9	1.7-1.2-1
Sewage sludge	3.5	2.5	.5	7-5-1
Cottonseed meal	6.5	3.0	1.5	4-2-1
Bat guano	13.0	5.0	2.0	6.5-2.5-1
Dried blood	13.0	1.5	—	—
Compost	1.5	1.5	1.5	1-1-1

*Figures shown are averages. Composition varies widely. Compost, for example, may analyze as high as 3.0% or as low as 0.7% in potassium content.
†P_2O_5
**K_2O

Organic fertilizers are low analysis fertilizers—a typical analysis would be 1-1-1 (Table 13).

Manure tends to have less phosphorus than nitrogen or potash, but this matches the way plants feed (see Table 12, page 46). Plant tissue contains less phosphorus than nitrogen or potassium, so we must assume that plants need less phosphorus. Don't worry about manure or other organic fertilizer being "weak" in phosphorus—nature has grown plants for millions of years with nothing but organic fertilizer.

Wood Ashes

Do wood ashes from a fireplace or other wood-burning stove make a good fertilizer? Yes, if you live where soils are naturally acid (pH 6.5 or less). Ashes are rich in potassium, and a good place to put them before they go into your garden is the compost heap.

Manure

Manure is one of the most common organic fertilizers, probably because it is produced in large amounts by farm and ranch animals and becomes a nuisance and a source of pollution unless disposed of.

The composition of manure varies widely. Feedlot manure is richer in plant food elements than most other manure. All manure (and probably most organic fertilizers) has a high residual value. That is, it lasts for years but the effect is greatest the first and second years, then decreasing until it is nearly all gone after about 10 years. About 45 percent of the plant food elements locked up in manure are released the first year, 22 percent the second year, and 11 percent the third year. Table 14 shows the residual effect of manure applications.

Let's take 100 square feet of garden soil and apply a ¼-inch layer of dairy manure (about 15 gallons) once a year for 8 years. Each yearly addition of manure builds up until the yearly release reaches 0.55 lb. of actual nitrogen, and levels off. The following years it increases very little.

When purchasing manure, be sure it has a low salt content (2,000 ppm or less). Some farmers mix salt in with their animals' food, which leads to high salt concentrations in the manure.

Applying manure as the only fertilizer. A light annual application of manure (¼ inch deep, or 15 to 20 gallons per 100 square feet) will furnish all the plant food elements your plants need (Table 15). At this rate and frequency, the levels of nitrogen would be maintained at about ½ lb. per 100 square feet per year. Levels of phosphorus and potassium would also be kept high at about ½ lb. (see footnote to Table 14). As a fringe benefit, all the other essential plant food elements will also be maintained in good supply.

When is "composted manure" not composted? To check to see if an organic fertilizer is really composted, take half a cupful, moisten it, place in a plastic bag, seal and place on top of your hot water heater. Check it after 3 days. If the smell is earthy, it is composted. If it stinks, it isn't.

Cottonseed Meal

Inside the seed of cotton is an oily kernel. The oil is pressed out for cottonseed oil and the seed coats are removed and sold as cottonseed hulls (good for improving soil structure). What is left is a yellowish cake high in protein. This cake is ground into cottonseed meal. It is an excellent organic fertilizer and is quite high in nitrogen (Table 13).

If you have a sick shrub or tree, try vertical banding of cottonseed meal under the drip line, using 1 to 2 cups per inch of trunk diameter and no

Table 14
Total Pounds of Plant Food Elements Released by Applying 15 Gallons of Manure per 100 Square Feet per Year*

	Year							
	1st	2nd	3rd	4th	5th	6th	7th	8th
1st Application	0.30	0.14	0.07	.039	.019	.009	.005	.003
2nd Application		.30	.14	.070	.039	.019	.009	.005
3rd Application			.30	.140	.070	.039	.019	.009
4th Application				.300	.140	.070	.039	.019
5th Application					.300	.140	.070	.039
6th Application						.300	.140	.070
7th Application							.300	.140
8th Application								.300
Total Lbs. Released:	.30	.44	.51	.55	.57	.58	.58	.58

*These figures are based on nitrogen release. Phosphorus and potassium release are very similar (about 93 percent of nitrogen).

Based on dry dairy manure with a nitrogen content of about 1.4 percent nitrogen.

The first year, 0.30 lbs. of nitrogen are released by the first application. The second year, the *second* application gives up 0.30 lbs. of nitrogen but the first application is also still on the job and releases 0.14 lbs., to give the total release shown at the bottom of the chart. The first application continues to release plant food elements for several years, the amount decreasing each year, but still adding some food.

Table 15
Effect of Manure Application on the Soil*

Depth of Layer Applied	Initial Increase in Organic Matter (%)	Effective† Increase in Organic Matter (%)	Manure Application	
			Lbs./100 sq. ft.	Gals./100 sq. ft.
¼″	1	.5	50	17
½″	2	.7	100	34
¾″	3	1.5	150	51
1″	4	2.0	200	68
2″	8	4.0	400	136
3″	12	6.0	800	204

*Based on dry dairy manure with a weight of about 600 lbs. = 1 cu. yd. and 1 gal. = about 3 lbs.

†Probable average organic matter content after initial rapid decomposition—3 to 12 months after application.

less than 1 to 2 cups per plant. I treated some sick rose bushes this way, and the cottonseed meal proved far superior to a popular slow-release pelleted fertilizer.

Plant Residues

Plant residues—non-woody prunings, old leaves, stems, or roots—can be composted or shredded and put back into your soil, or on it as an organic mulch.

I never cease to be amazed at the vast amounts of organic matter going to the dump from our towns and cities. It is a great waste of a most valuable natural resource. A wise gardener will keep a covered pail in the kitchen in which to place non-mushy garbage, coffee grounds, vacuum cleaner sweepings, hair trimmings and other organic matter that can go into the compost heap.

How to Apply Fertilizers

Annuals—plants like corn or beans or marigolds—are replanted every year, so the soil can be worked up and fertilizers mixed in.

With perennials, such as asparagus, berries, shrubs or lawns, you cannot work up the soil to mix in fertilizer. For these perennials, fertilizer must be applied as a *topdressing.* This means you broadcast it on the surface and water it in. Sometimes it is also scratched lightly into the surface and watered in or left for the rain to carry downward.

"Banding" Phosphate

Phosphate is best applied to soil as a band under the seed rather than broadcasting and mixing in. The broadcast-and-mix-in method is inefficient because the phosphate reacts with the soil to form non-available phosphorus compounds.

One study showed that it took 200 lbs. per acre of phosphate mixed in the soil to equal the effect of 30 lbs. per acre banded. In a study done in the Rio Grande Valley of Texas, yields of tomatoes were increased from 2 tons per acre to 11 tons per acre when phosphate was banded instead of broadcast. The treatment was equal to about ¼ cup (about 2 oz.) of superphosphate (0-20-0) per plant placed 2 to 3 inches directly under the seed (Figure 20).

Try this on a few of your tomato plants (be sure to leave some untreated for a good test). If you

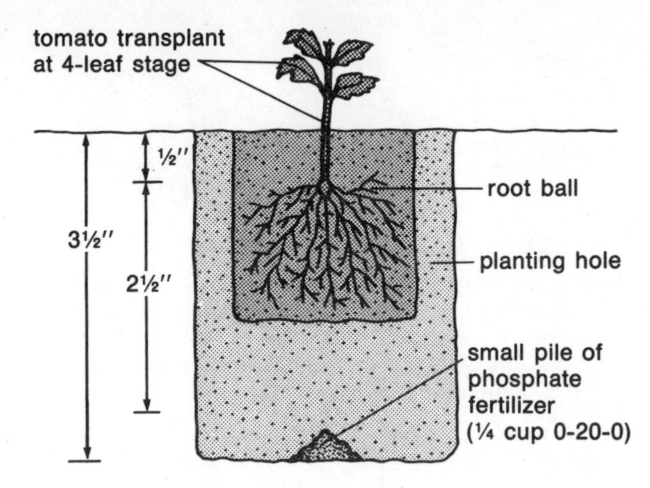

Figure 20. *Super tomato plants by fertilizing with superphosphate.*

start with the tomato plant, assume the seed was planted ½ inch below the soil line, so dig your planting hole 3½ inches deep and pour ¼ cup of superphosphate in a pile in the bottom of the hole. Cover the pile of phosphate with about an inch of good soil or compost, then set in your tomato plant (Figure 20). Fill in the sides about halfway up with good soil and pour in a pint or so of water with a pinch of superphosphate dissolved in it. This will settle the soil and destroy air pockets. Then finish filling in the sides, firm soil to form a small cavity, then pour ½ pint or so of the fertilizer solution in the cavity. When the water has soaked in, rake some loose soil in to fill the cavity.

Will the above method work with pepper, okra, beans, peas and other vegetable crops? I have no data to prove it, but phosphorus promotes seed production, which also promotes growth of the fruit surrounding the seed. So it should work for almost all fruit-producing plants.

What Makes Banding Work?

To keep phosphorus (and potassium, to some extent) "alive," active, and available to the plant, the fertilizer particles must somehow be protected. The best way to do this is to make sure they "stick together." By lying in a closely-knit mass, they may sacrifice some of their outer members, but the core remains protected and intact (Figure 21). If scattered and mixed with the soil, all particles are soon neutralized (Figure 22).

Banding a Complete Fertilizer

A *complete* fertilizer is one that contains all three of the "Big 3" elements—nitrogen, phos-

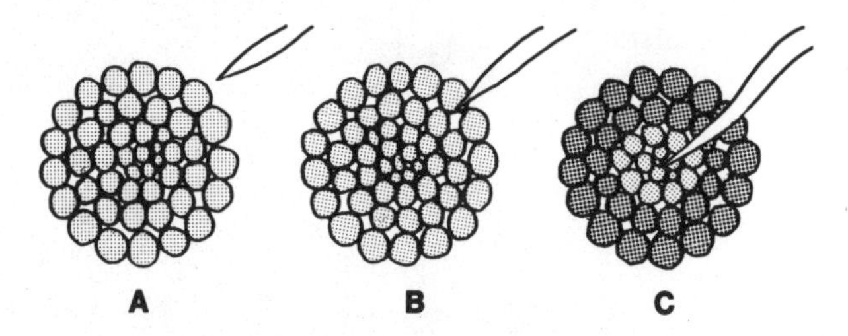

Figure 21. Keeping phosphorus available to plant roots by "banding." A, B—Root approaches and begins feeding on a band of phosphate. C—A month later, much of the phosphate is neutralized (darker areas), but the inner core is still active and useful to plant roots that penetrate and feed.

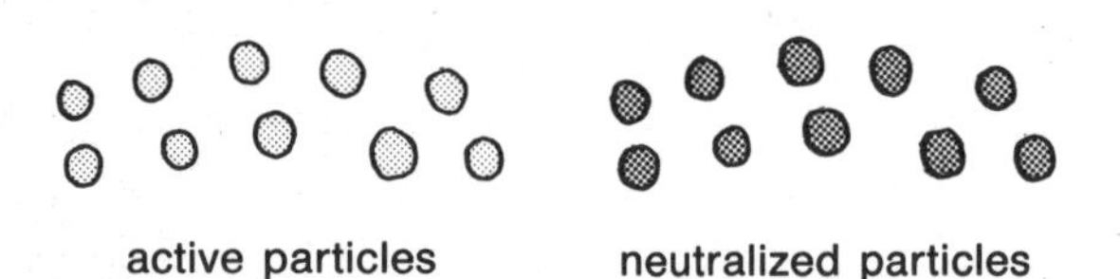

Figure 22. When you broadcast a phosphate fertilizer on top of the soil, reactions with other chemicals "neutralize" or deactivate all of the thinly scattered particles.

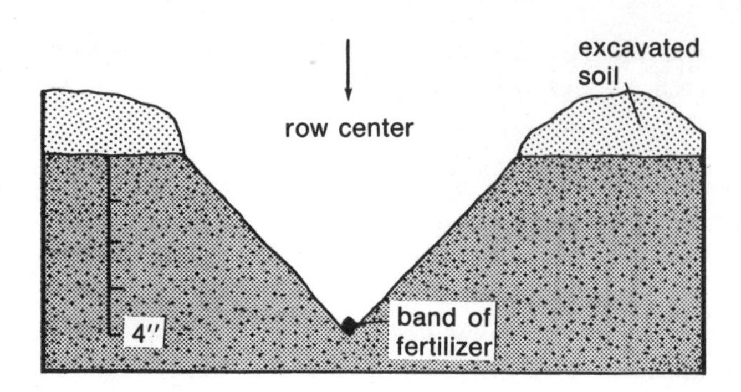

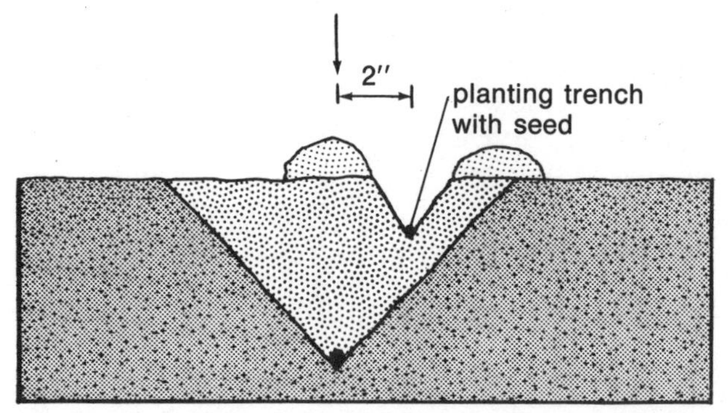

Figure 23. Applying a band of complete fertilizer at planting time. Place the band of fertilizer 2 or 3 inches beneath the seed of the plants you intend to grow.

phorus, and potassium. If you broadcast and mix in a complete fertilizer, most of the phosphorus will be neutralized.

You could band all three below and to the side of the seed (Figure 23). But excess rainfall or too much irrigation may leach away much of the nitrogen before the plants get big enough to really use them.

Perhaps the best approach is to apply banded phosphate at or just before planting at the rate of about 1 to 2 lbs. of superphosphate (0-20-0) per 100 linear feet of row. Apply 2 to 3 inches directly under the seed or under the plant. Then, when the plants are up with about 4 true leaves, sidedress them with a balanced fertilizer.

How much fertilizer should you use when banding? It depends on how high the percent of nitrogen (Table 16). For any crop that requires over 45 days to mature, a *split application* is probably best. A split application means you split the total amount of fertilizer into two halves and apply the first half when the plants have 2 to 4 true leaves and the second half about a month later.

Table 16
Amount of Balanced Fertilizer To Apply When Sidedressing*

Nitrogen (%)	Total lbs. per 100 linear ft.	lbs. to Apply: One Application Only	lbs. to Apply: Two Applications
5	3	3	1.5
10	1.5	1.5	0.75
15	1	1	0.5
20	0.75	0.7	0.35

*A balanced fertilizer has all three numbers written on the bag, with none of the numbers being a zero; for example, 5-10-5 or 8-8-8 or 12-24-12.

Applying Nitrogen Fertilizer

If your soil test should come back with phosphorus and potassium "High" or "Very High," chances are the nitrogen level will be "Low." This is more likely to occur in areas of lower rainfall, where minerals have accumulated. In such areas nitrogen may be the only thing you can get a response from. So, wait until you have a good stand of plants to utilize the nitrogen, then apply it as a sidedress or topdress and water in.

Sidedressing with Nitrogen

Once the supplies of phosphorus and potassium have been built up in the soil, nitrogen can easily be added by broadcasting on the surface over the root zone and watering in (or applying just before a rain). You can also apply it as a sidedressing if you don't get too close to the row of plants (Figure 24).

Follow these guidelines:

1. Don't apply within a month of maturity, otherwise maturity may be delayed.
2. Apply ½ to ¾ lb. of ammonium sulfate (21-0-0) or equivalent per 100 linear feet of row and water in.

A good way to apply ammonium sulfate is to use a hose-on applicator and a 1 percent solution (1 lb. of fertilizer per 12 gallons of water). Or you can make a shallow trench alongside your row of plants with a hoe, mix the pound of fertilizer in 12 gallons of water in a garbage can, then apply the fertilizer solution with a sprinkler can with the sprinkler head taken off. You can make a concentrated liquid nitrogen solution by mixing 1½ lbs. of 21-0-0 in a quart of water. Then 1½ cups of this per 100 linear feet of row applied in 12 gallons of water comes out just right.

Dry, granular fertilizer can also be mixed with water to make liquid fertilizer (Table 17).

Nitrogen Helps Phosphorus Uptake

If a mixture of nitrogen and phosphorus is applied in a band, the nitrogen stimulates plants to absorb phosphorus. Instead of banding just 0-20-0 (superphosphate), it may be better to band 16-20-0 (ammonium phosphate). The exceptions are soils in low rainfall areas that may be so high in phosphorus already that no response occurs when additional phosphorus is applied.

Tree Feeding

Proper tree feeding begins with the planting of a baby tree—plant it level with the soil surface, not in a sunken basin. The sunken basin works fine the

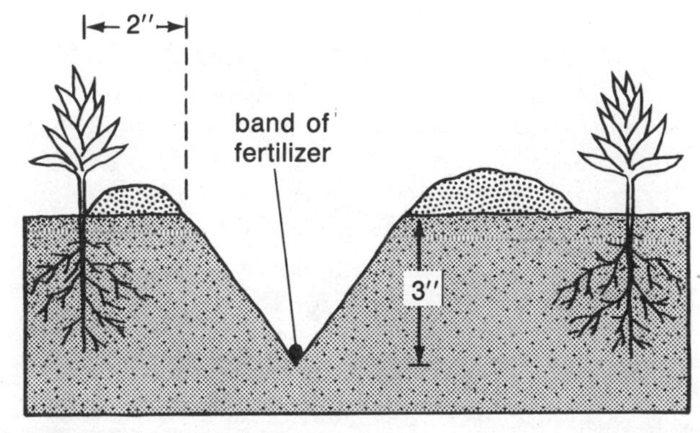

Figure 24. Side dressing a row of young plants with a band of complete fertilizer.

Table 17
Amounts of Salt-Type Fertilizers Dissolving in One Gallon of Water

Fertilizer	Approx. lbs. Soluble in 1 gal. of Tap Water
Ammonium nitrate	9.8
Ammonium sulfate	5.9
Calcium nitrate	8.5
Diammonium phosphate	3.5
Monoammonium phosphate	1.9
Sodium nitrate	6.0
Potassium nitrate	1.0
Superphosphate, single	0.15
Superphosphate, treble	0.33
Urea	6.5

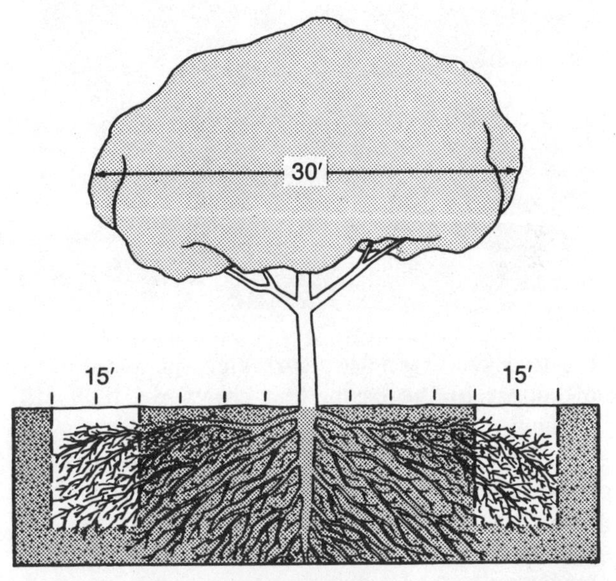

Figure 25. The areas underneath the drip line of a plant are the zones of greatest root activity—plants do 70 percent of their feeding in this zone.

first year or two but later is a curse because the water collects in the *wrong* place.

An older tree feeds mainly from a ring of feeder roots centered under the drip line of the tree (the outer periphery of leaves). These roots do 65 to 70 percent of their feeding from a band 3 to 6 feet wide under the drip line (Figure 25). A 10-year-old tree will extract about 66 percent of its water from this band and only 5 percent from a 6-foot circle around the trunk. Since plants cannot feed on fertilizer until it gets dissolved in water, wherever the water goes, so goes the nitrogen fertilizer.

A young tree watered in its original sunken basin 3 to 4 feet in diameter around the trunk is locked into poor growth. If you fill in the basin, rot may develop and kill the tree. And it's always tempting to just drop the hose in the basin to ir-

A fertilizer spike. You simply drive these solid hunks of fertilizer down into the soil.

rigate. Even if you use a sprinkler, the water tends to collect in the basin, and crown rot may still develop.

Always plant a tree with the soil line on the tree trunk even with the natural soil surface and depend on a sprinkler or rainfall for watering. You can broadcast organic fertilizers around the active feeding ring and water in.

Subsurface Root Feeding

This is often done by inserting a long, slender, hollow pipe down into the soil where the roots are and injecting a fertilizer solution under pressure. It is advantageous when:

1. Top layers of soil are tight and water does not penetrate easily so that you cannot broadcast nitrogen fertilizer on the surface and water in.
2. You don't want to go to the trouble of watering fertilizer in. In subsurface root feeding the fertilizer is mixed with water before injection. Professional tree care people often use subsurface root feeding because they can apply fertilizer and water at the same time and know the entire job is done correctly.
3. You're applying phosphate and potash as well as nitrogen. If you broadcast potash and phosphate on the surface, that's where they are likely to stay; they won't get down into the top foot of soil, where most of the feeding occurs.

Subsurface feeding is usually used for trees and shrubs. There are four methods: (1) pressure injection of liquid fertilizer, (2) poking holes and pouring in a water-and-fertilizer solution, (3) poking holes and pouring in dry granular fertilizer, or (4) driving solid spikes of fertilizer down into the soil.

Trees will feed most in the top 1 to 2 feet of soil. If the tree is in a sodded area, the grass root feeding zone is likely to be the top 8 inches. So, if you have grass around the tree, between 8 and 18 inches is the best depth.

If the fertilizer you're applying contains phosphate, the phosphorus in the phosphate will likely be neutralized by soil reactions—if you apply it in liquid form (see page 49). So, it's best to use the dry granular fertilizer or drive in tree spikes. In essence, this is the same as banding the fertilizer under rows of the garden, but now you're laying the fertilizer in *vertical* bands for the deep tree roots.

Vertical Banding

Under the drip line, make a ring of holes 16 to 18 inches deep, ¾ to 1 inch in diameter, and about 36 inches apart (a pace apart is about right). Let the water subside before pouring in the dry fertilizer.

Use the tree's trunk diameter to calculate the amount of fertilizer needed. You can use a tape measure or flexible rule to measure the distance around the trunk (the circumference), then divide this figure by 3 to get the diameter. A rule of thumb is 1/10 lb. of *actual* nitrogen per inch of trunk diameter. (If you're using fertilizer spikes, two spikes per inch of trunk diameter will be sufficient.)

Fertilize just before or about the same time that growth begins in the spring. Use the first number on the fertilizer bag (percent nitrogen) in computing amount needed; the higher the nitrogen content, the less you use (Table 18). Measure the trunk diameter, calculate the number of pounds of fertilizer needed, and measure it out in a bucket.

When applying, divide up the fertilizer so that each hole under the drip line gets about the same amount. A large-neck funnel is handy for getting the fertilizer down in the holes. Fill the holes with fertilizer, making bands about 8 inches long (Figure 26). Then finish filling the top part of the empty hole with a sand and soil mixture. Finally, water the entire area thoroughly.

If you want even more growth, make a second application around June 1.

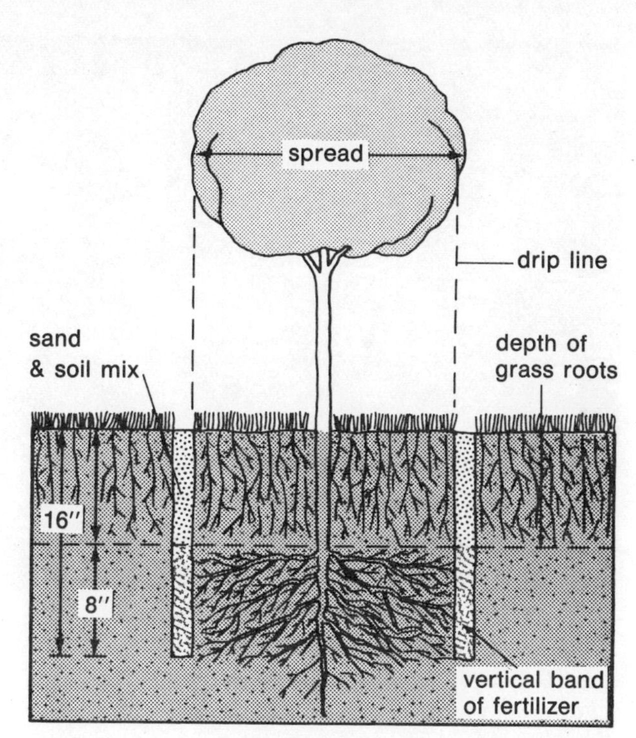

Figure 26. *Applying a vertical band of fertilizer to tree roots. Holes are about 36 inches apart around the tree and about 16 inches deep. (Not to scale.)*

Fertilizing newly planted trees. If you're planting a container plant, it's all right to fertilize at planting. If it is a bare-root baby tree (or any other bare-root tree), don't feed the first year, unless a deficiency develops. Wait a year for roots to first develop. Newly planted pecan trees are especially sensitive to fertilizer and may be killed.

Table 18
Fertilizing Fruit and Nut Trees with Nitrogen Fertilizers

Nitrogen in Fertilizer (90)	Amount of Fertilizer to Supply 0.1 lb. of Nitrogen*	
	Weight	Cups
21	7 oz.	¾ rounded
20	8 oz.	1
19	9 oz.	1 rounded
18	10 oz.	1¼
17	10 oz.	1¼
16	11 oz.	1¼ rounded
15	12 oz.	1½
14	13 oz.	1½ rounded
13	14 oz.	1¾
12	14 oz.	1¾
11	15 oz.	1¾ rounded
10	1 lb.	2
9	1 lb. 2 oz.	2¼
8	1 lb. 5 oz.	2½ rounded
7	1 lb. 7 oz.	2¾ rounded
6	1 lb. 11 oz.	3½
5	2 lbs.	4

*Based on dry, granular fertilizer (salt-type), where 1 pint = about 1 lb. Check your fertilizer to see if this agrees. All cups are level unless otherwise shown.

Trace Elements in Vertical Bands

Some plants have a great deal of trouble absorbing enough trace elements for proper nutrition. The pecan tree is a prime example—it often cannot get enough zinc. Many plants—peach trees, for example—have iron deficiency, especially in low rainfall areas that have a high soil pH. The vertical banding fertilization just described is effective in preventing both zinc and iron deficiencies.

Foliar sprays are also effective, but they're a lot of work and only temporary in effect. And if you hire someone to spray a huge pecan with zinc the required three to five times during the season, you'll need to sell some pecans to replenish your bank account.

How much zinc and iron? For pecans, 1 to 1½ lbs. of zinc sulfate per inch of trunk diameter is effective in prevention of zinc deficiency.

For peaches and other trees suffering from iron hunger, 1 lb. of iron sulfate per inch of trunk

The influence of zinc sprays on pecan leaf size: no supplemental zinc (left); with zinc sprays (right).

diameter is adequate. Don't feed a newly planted tree the first year. A starter solution of phosphorus at planting is all right, but no more than this.

Feeding Your Lawn

Feeding a lawn with salt-type fertilizer is really two jobs in one: (1) broadcasting the pellet-like granules of fertilizer, and (2) watering it in. It won't work unless you water it in, and it may burn the lawn if you don't.

To water fertilizer into the root zone of a lawn, use a sprinkler to apply about ½ inch of water—unless you can get just ahead of a rain. To determine when ½-inch of water has been applied, place an empty coffee can about 4 feet away from the sprinkler, set your kitchen timer for 15 minutes, then go out with a ruler and measure the depth of water caught.

A good lawn feeding guideline is 1 lb. of actual nitrogen per 1,000 square feet of area. (This would be 5 lbs. of 21-0-0 per 1,000 square feet. You can use Table 18 to find the amount of nitrogen fertilizer needed, but multiplying the amounts shown by 10 (Table 18 shows the amounts needed for 1/10 lb. of actual nitrogen).

Complete Fertilizers and Lawn Feeding

It's no problem to get nitrogen down into the root zone of grass where it can be used, but phosphate and potash are a different story. Most turfgrass specialists recommend that once or twice a year a complete fertilizer be applied to the lawn.

Apparently, the phosphorus, and especially the potassium, go deep enough into the soil since good response has been reported when final fall feeding is done with a complete fertilizer. This is especially true when the fertilizer has a high potassium content, which builds cold-resistance into the plant.

An ideal combination for fall "winterizing" is nitrogen and potassium. A 16-0-8 or 12-4-8 analysis is perfect.

Fertilizing a Lawn Before Planting

If the soil is low in phosphorus, apply it before planting. Though phosphorus works best in bands, it's not feasible to band it because there are no rows of plants. The next best thing is to apply it on the surface and mix in with a rototiller.

It is best *not* to apply a complete fertilizer at planting for two reasons: (1) weed seeds tend to come up first, and nitrogen just stimulates them to grow faster and compete with the lawn grass you have seeded, and (2) if heavy rains or over-irrigation occur before grass seed germinate and emerge, the nitrogen may be leached out of the soil and lost.

Wait until you have a good stand of grass, then topdress with nitrogen-only fertilizer. As soon as you have a good stand of grass seedlings broadcast 5 lbs. of 21-0-0 per 1,000 square feet when the young seedlings are first seen, water in, then come back a week later and repeat the first feeding.

As Dr. Duble recommends in his book, *Southern Lawns and Groundcovers* (Pacesetter Press, Houston, Tx.), it is good to broadcast and rototill in limestone at the rate of 25-50 lbs. per 1,000 square feet if your soil is acid (pH under 6.5); if the pH is very alkaline (8.5 or more), 50-100 lbs. of gypsum per 1,000 square feet rototilled in will help.

Fertilizer Conversion Charts

Conversion from Pounds Per Acre Into Weights for Small Areas

Rates per acre (lbs)	Lbs per 1000 sq ft	Lbs per 100 sq ft
100	2½	¼
200	5	½
400	9	1
800	18½	2
1000	23	2½
Manure, leaves and straw		
8000 (4 tons)	200	20
16000 (8 tons)	400	40

Conversion from Area Rate to Linear Row

Rates per 1000 sq ft	Rates per 100 sq ft	Row width 3 ft	Row width 2 ft	Row width 1 ft
Lbs	Lbs	Lbs per 100 ft of row		
5	½	1½	1	½
10	1	3	2	1
20	2	6	4	2
30	3	9	6	3
40	4	12	8	4
50	5	15	10	5

Factors for Converting Suggested Rates When Substituting A Different Analysis Fertilizer

Suggested analysis	Analysis to be substituted	Rate conversion factor*
12-12-12	20-20-20	.6
	17-17-17	.7
	16-6-12	.8
	15-15-15	.8
	13-13-13	1
	12-6-6	1
	10-5-5	1.2
	10-6-4	1.2
10-20-10	5-10-5	2
	5-15-5	2
	6-12-6	1.7
	12-24-12	.85
	15-30-15	.65
10-20-20	12-24-24	.85
	15-30-30	.65

*Multiply times rate for suggested analysis. Example: want to substitute 20-20-20 for 15 pounds 12-12-12. 15 x .6 = 9 pounds 20-20-20 needed.

Conversion from Area Rate to Per Plant Basis

Rates per 100 sq ft	Spacings 5 x 5 ft	Spacings 2½ x 2½ ft	Spacings 2½ x 1½ ft
Lbs	Oz * per plant		
½	2	1	½
1	4	2	1
2	8	4	2
3	12	6	3
4	16	8	4
5	20	10	5

*1 oz. = 2 tbsp.

Special Fertilizers

Starter Solutions

When a plant is transplanted, either bare-root or with a root ball, transplant shock may occur. The reason for this is that the transplant has a limited root system and thus a limited capacity to take up plant foods and water. Adding a starter solution of liquid fertilizer at transplanting time will reduce shock and help the plant get off to a good start.

A starter solution should be rich in phosphorus, since this element stimulates root formation and early growth. It also has a low salt index and won't injure young roots. Another important rule is to keep the solution weak.

Probably the best fertilizer to use in making a starter solution is superphosphate (0-20-0). Use 2 tablespoons of 0-20-0 or 1 tablespoon of 0-46-0 per gallon of water and stir well to mix. Pour a pint of solution around each small transplant at transplanting. For larger plants, like trees and shrubs, up to 5 gallons of solution may be needed.

Iron Fertilizers

If your soil pH is 6.5 or less, you probably won't have iron deficiency. If you do, a leaf spray of ferrous sulfate and water will help. Mix rounded tablespoonsful of ferrous sulfate with water as follows:

Fruits: 1 tablespoonful per gallon
Ornamentals: 2 tablespoonsful per gallon
Vegetables: 3 tablespoonsful per gallon
(Add ¼ teaspoon of liquid soap per gallon as a spreader)

Instead of a leaf spray, you can sidedress with liquid iron by mixing ¼ cup of ferrous sulfate per gallon of water. Make a trench and apply 25 gallons of solution per 100 linear feet.

If your soil pH is 6.6 to 7.0, you can get better results using iron *chelate.* Iron chelate is a "bonded" form of iron; the iron will remain available to the plants when it is supplied in this form, and this is often necessary with more alkaline soils. (See the iron bar on the pH chart, page 8.)

A chelate called "Sequestrene Fe 330" should work if mixed with water and applied as a soil drench. If you have a soil pH of 7.5 or over, the only good iron chelate for soil application is "Fe 138." This is a light, fluffy, blood-red powder containing 6 percent iron. One oz. equals 3 to 4 level tablespoonsful. Fe 138 is hard to mix with water if you don't know how. To do it easily, get a 1-gallon wide-mouth jug with a tight-fitting lid. Place 1 cup (6 oz.) of the red powder in the jug. Add 6 cups of water. Shake 30 seconds to mix. Then pour the red liquid into an empty 1-gallon jug. Label this jug "1 oz. iron chelate per cup."

You now have a liquid iron concentrate. Every cup contains 1 oz. of the red powder. For a soil drench, add 1 cup of your liquid iron to a gallon of water. Use 1 to 3 oz. of red powder per inch of trunk diameter or ½ to 1 oz. per 100 square feet of flower bed. Read the label for further directions.

Timed-Release "Plant Food"

These may be referred to as "time-release," "slow-release," or "controlled-release" fertilizers. They consist of fertilizer particles with a coating of some other material that prevents the fertilizer particle from dissolving in water immediately. Instead, the nutrients are released slowly over a long period of time, usually 3 to 6 months.

These salt-type man-made fertilizers are in sharp contrast to the regular salt-type fertilizers that burn, leach out easily, and kill plants if overapplied. They are more like the natural organic fertilizers, but there is still one major difference: natural organic fertilizers are not only slow-release but also contain 13 essential plant food elements: the man-made slow-release fertilizers contain mostly nitrogen, phosphorus, potassium, and sometimes sulfur.

Examples of the slow-release fertilizers are "MagAmp" or "Osmocote," with 14-14-14 analyses. I have tested "Osmocote" with good results in growing container tomatoes in a half sand-half peat moss mix.

A problem with most of these slow-release fertilizers is that they lack the other ten essential elements plants must have. An easy way to solve this problem is to mix in a little manure (see page 49 for details). Or, you can mix in some of the other organic fertilizers to supply the trace elements.

Free Fertilizer from the Atmosphere

Plants must have nitrogen—and lots of it. The atmosphere surrounding our planet is mostly nitrogen in the form of a colorless, odorless, tasteless gas. In fact, about 80 percent of air is nitrogen. The trick is to somehow capture the nitrogen and put it into a form plants can use. Fertilizer factories use atmospheric nitrogen as their source when manufacturing nitrogen fertilizer.

An inoculant for legumes such as peas, beans, and peanuts. The inoculant supplies the bacteria that work with the plants' roots to get nitrogen in the air down into the soil.

Nature uses the same source, but does so biologically. The organisms that do it are called "nitrogen-fixing" bacteria. There are two of these. One, *Azotobacter*, lives in the soil as a "free-living" bacteria. The other, *Rhizobium*, doesn't like to live free and works only when attached to the roots of certain plants called legumes. Legumes are plants like beans, peas, and peanuts. When the roots of a legume grow down into the soil, the *Rhizobium* bacteria attach themselves to the roots and set up a partnership with them—to everyone's benefit. A small, wart-like nodule develops as a biological, nitrogen-fixing factory (don't confuse these nodules with root knot nematodes; *Rhizobium* nodules are much smaller and when cut open usually contain an amber-colored fluid).

How do you get these biological nitrogen-fixing factories to work for you? First, you have to plant a legume, then see if it develops the nodules. If it doesn't, your soil may be too rich in nitrogen for the bacteria to work or there may be no bacteria present. The easiest way to get them in your soil is to let the seed carry them there. You buy a small bag of sooty, black powder that contains the bacteria. It is called inoculant, and be sure to get the right kind to fit the kind of legume you are going to plant.

Moisten the seed and sprinkle the black powder over the moist seed as you stir. Peanuts are especially responsive to seed inoculation, particularly in sandy soils.

Friends of the Earth: Earthworms and Mulches

On the face of it, soil seems a lifeless mass of tiny rock particles with a little water and plant food. A good, fertile soil is far more: it is almost alive—abounding with millions of living organisms. These may be microscopic plants such as bacteria and yeast-like cells or beneficial animals like ants and earthworms. An acre of fertile soil may contain 1,100 lbs. of live worms, or 2½ lbs. per 1,000 square feet!

The Worm Cure

Once, when I was a boy, my mother had a hanging basket of asparagus fern that had seen better days. It was alive, but barely, and I couldn't help but feel sorry for it. One day I read about earthworms and how they help soil, so I went out and dug one up and put it in the hanging basket. No other changes were made. After about a week it looked as if springtime had come—the sick fern had taken on new life, sending out new shoots everywhere. On the surface of the soil I saw several mounds of worm castings. In a few weeks the old, sick fern had been transformed into a luxuriant thing of beauty, all due to one small earthworm.

You may doubt what you hear; but what you are able to demonstrate, you never forget. To this day I have an abiding respect for earthworms and what they can do for soil fertility and plant growth.

Are Earthworms Really Beneficial?

It's a common notion that the number of earthworms in a soil indicates the soil's fertility. But on the other hand, you *can* grow bumper crops of nutritious vegetables and other plants without a single earthworm. So, do earthworms make soils fertile, or do fertile soils attract earthworms?

The answer is yes, and yes: earthworms are very fond of soils rich in organic matter (too much man-made fertilizer will harm them, though), and they benefit soils by aerating them and by redistributing organic matter.

How the Earthworm Lives

The earthworm is technically known as a roundworm. It has both male and female (hermaphroditic) sexual characteristics. Reproduction is by eggs.

The skin of the earthworm must be constantly moist or it will die. This is why earthworms are not found in coarse, sandy soils unless there is a heavy sod or a thick mulch. Earthworms also need a soil with good supplies of replaceable calcium. Where soil pH is under 6.0, liming will encourage them.

A mature earthworm generally weighs about half a gram. Earthworms deposit "eggs" which are really small brown egg cases about ⅛ inch in diameter (about the size of a grain of wheat). Each egg case contains 3 to 10 eggs. Young worms hatch in about a month and are full grown in 2 to 3 months.

Earthworms feed mostly at night. They have no eyes but are sensitive to light.

Strong salt-type fertilizer may discourage worms if heavy applications are made. The saltier the fertilizer, the less worms seem to like it (see Table 11, page 45). Ammonium sulfate, though it is a useful fertilizer, is a product that does not encourage earthworms.

How Earthworms Make Soil Fertile

Earthworms eat soil to get the organic matter in it. They don't do this by choice, but by necessity. If they could somehow extract the organic matter from soil without having to eat the soil, they probably would.

They also eat soil in order to make their burrows. They swallow the soil, it passes through their digestive tracts, then the worm crawls up to the surface and excretes the digested soil as worm castings. He has "killed two birds with one stone"—he has fed himself and at the same time dug a tunnel for his home.

It is estimated that the earthworms in an acre of fertile soil will eat and pass through their bodies up to 30,000 lbs. of dry soil per year. This amounts to about 11 oz. of dry soil per square foot each year, or a layer about 1/10 inch deep.

The small pile of particles alongside a hole occupied by an earthworm is a pile of "castings," or worm manure. But it is different from any other animal manure we know of. It is odorless and "clean."

In digesting the soil the worm grinds it and exposes it to enzymes. The worm excretes the soil, minus some of the organic matter, as castings. These castings act like fertilizer on plant growth.

Even though earthworms extract organic matter from the soil they eat, they apparently put some back. It is a fact that organic matter, phosphorus, potash, and general availability of soil nutrients are all increased through earthworm activity.

Earthworms also benefit the *physical* nature of the soil. When you poke a hole in the soil, you increase aeration by means of the tunnel you have just made. Earthworms do the same thing, only better: the holes they make are smaller and more numerous.

With a rototiller you mix organic matter and surface soil with the deeper soil layers. Earthworms do a better job of this, too: they dig small tunnels through the soil and drag raw organic matter (food) down into them. All this brings about increased pore space, soil granulation and better structure, all described earlier in Table 1.

How to Encourage Earthworms

Become familiar with their nature and requirements. Avoid the things that injure them. Keep the soil supplied with natural organic matter and never let the top foot of soil dry out. A surface mulch is very helpful in this regard. Farmyard manure also has a pronounced beneficial effect on earthworm populations.

In summary, to encourage earthworms:

1. Maintain high organic matter content in soil.
2. Mulch the soil surface.
3. Irrigate, if necessary, to keep the top foot of soil moist.
4. Keep soil pH 6.0 to 7.5 if possible.
5. Maintain good subsurface drainage (avoid waterlogged soils).
6. Avoid heavy applications of salt-type fertilizers such as ammonium sulfate.

Earthworm Compost

Earthworm "compost" is organic matter that has been worked over and run through the digestive tracts of earthworms and that still contains many worms. This compost is prepared in special worm beds.

Long, shallow bins sitting on top of the ground about 1 foot deep by 5 feet wide by 20 feet long are

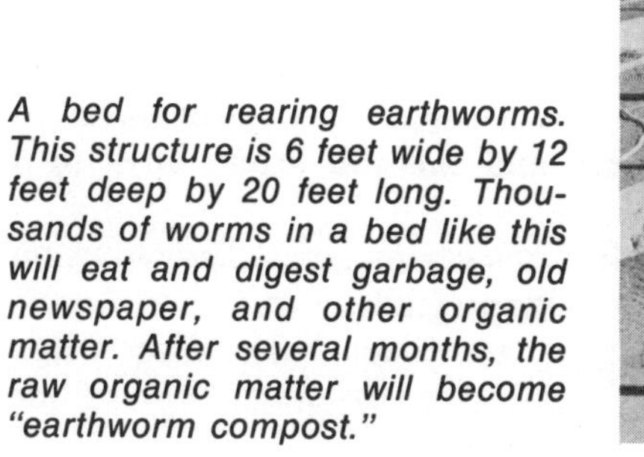

A bed for rearing earthworms. This structure is 6 feet wide by 12 feet deep by 20 feet long. Thousands of worms in a bed like this will eat and digest garbage, old newspaper, and other organic matter. After several months, the raw organic matter will become "earthworm compost."

A handful of earthworm compost teaming with earthworms. Little else is as beneficial to the soil as this material.

often used as the worm rearing beds. Shredded newspaper from old newspapers run through a compost grinder makes good food for worms and may be placed on top of the beds and wet down. After a month or two, raw garbage, manure, leaves and other organic matter will have turned into rich compost teeming with thousands of earthworms.

Sometimes you can buy this material by the cubic yard just for the compost value. If the price is $27 a cubic yard (about 200 gallons), you'll probably find this cheaper than buying a bale of peat moss at a garden center. Often, common types of organic matter will sell for $2 per cubic foot or $54 per cubic yard. If you have earthworm farms in your vicinity, you may be able to get this high-quality worm compost at a reasonable price. Call your county Extension agent for sources.

Other Beneficial Soil Life

Soil fungi are microscopic plants that feed upon soil organic matter. One helpful soil fungus is *mycorrhizae.* Oddly enough, like harmful fungi it invades the cells of plant roots, but instead of hurting the plant, helps it. Harmful fungi are takers; *mycorrhizae* take *and* give, and they give more than they take, so the plant comes out ahead. The fungus sets up a mutually beneficial partnership with the plant, helping the roots absorb more plant foods so it can have some too.

Soil bacteria are another form of beneficial microscopic plants in the soil. Fertile soils contain more bacteria than poor soils: a fertile soil may contain 1,000 lbs. of bacteria per acre 6 inches deep. Bacteria are necessary to transform organic matter into plant foods. Two bacteria—*Azotobacter* and *Rhizobium*—help furnish nitrogen to plants (see pages 56-57).

Mulches and Mulching

A mulch is a covering placed over the soil surface. It serves three purposes: (1) to control weeds by depriving them of light, (2) to conserve soil moisture by cutting down on surface evaporation, and (3) to lessen the fluctuations in soil temperature of the top 2 to 3 inches of your garden soil.

Put another way, a mulch on the surface of your garden soil inhibits the growth of all plants except the chosen crop. It reduces soil moisture loss from evaporation. It protects plant roots (and earthworms) from heat and cold. It helps keep fruit (such as strawberries) clean.

Mulches may be organic (straw, grass clippings) or synthetic (black plastic, fiberglass).

Some mulches not only control weeds and conserve soil moisture, but also allow air and water to pass through while blocking light. These are the best kind of mulches and include both organic mulches and non-organic types such as fiberglass. The plastic sheet mulches block light but they also block air and water movement into the soil.

Fiberglass Mulch

Fiberglass mulch is a synthetic mulch that is about ¼ inch thick and comes in rolls 150 feet long and in widths of 24, 36, 48, and 72 inches. The home vegetable gardener will find the 24- or 36-inch width the most useful.

Fiberglass mulch is porous, so it allows air and water to pass through but blocks light and retards loss of soil moisture. Steel pins are available to anchor it to the ground. It is especially useful for shrub borders where it can cover the surface in place of coarse bark. It is rot-proof, vermin-proof, and fire-proof. For pot plants, it comes (or can be cut) in circles and is installed on top of the pot to stop surface moisture loss.

Fruit, Nut, and Shade Tree Mulching

Fiberglass mulch is especially useful for mulching young trees. Using the 6-foot width, cut off a piece 6 feet long. This is made into a "mulch collar" (Figure 27) by cutting a circle out in the center (about the same diameter as the tree trunk) and then cutting straight out to the edge to make it possible to place the collar around the tree trunk.

Organic mulches can also be used with good results around trees, shrubs, rose bushes, etc.

Black Plastic Sheets

Plastic sheeting may be used as a mulch when growing widely spaced plants like tomatoes or

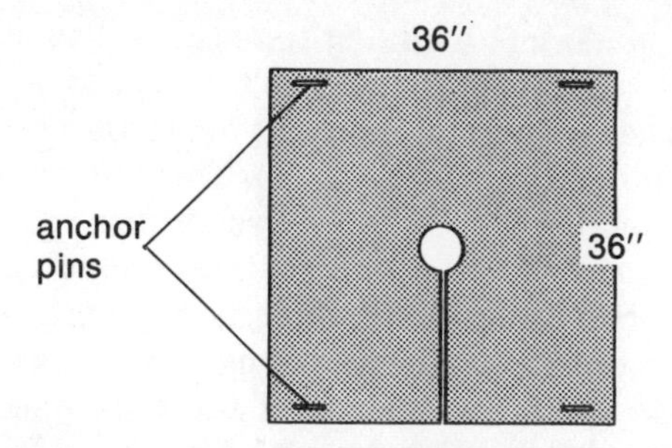

Figure 27. A homemade mulch collar.

melons. (Use black because it blocks light best.) The most useful width is probably 36 inches.

After preparing the seedbed, install the plastic sheeting and anchor it at the edges with soil. Along the centerline, punch circular holes and insert transplants in the holes. Usually, only the widely-spaced plants are planted this way. These include tomatoes, peppers, eggplants and melons. This mulch can be used for close-spaced plants like beans, corn, and radishes, but instead of circular holes along the center of the plastic, make slits by cutting a straight line 6 inches long, skip 2 inches, cut 6 inches, and so on. This provides an opening for young plants to push up through. The slit method is probably more trouble than it is worth; plastic sheeting is best used with plants spaced 1 to 2 feet or more.

Problems with plastic sheeting are (1) in windy areas it may be ripped up, and (2) this mulch will not allow rain to pass through. You also have the problem of removing it at the end of the season.

Black plastic has a warming effect on the soil, so it's good to use in getting a crop started earlier in the year.

Clear Plastic Mulch

Instead of black plastic, clear (transparent) plastic sheeting may be used. It warms up the soil more than black plastic, but it doesn't block out light, so weed seeds will germinate underneath it.

Rock Mulch

Medium-size gravel is useful as mulch around perennial plants such as trees and shrubs. To block out light, a layer 2 inches thick is necessary. Where plants are not to be grown in the mulched area, plastic sheeting is usually laid down, then the gravel placed on top. But where plants are to be grown, it's desirable to have water and fertilizer penetrate down into the mulched soil below. White gravel will reflect heat, black gravel or rock will absorb it.

Paper Mulch

Old newspaper can be used as a mulch. Open the folded sheets and make them 4 to 6 sheets thick. Soak in water and press flat against a level surface (first shave off any weed growth with a hoe and make the surface level). Use your ingenuity to find a way of securing the paper to the soil.

Paper mulch has a cooling effect on soil since the whiteness of the paper reflects the sun. The best time to apply it is after plants are several inches or more in height. Then the newspaper can be laid in the aisles between the rows.

Erosion Netting

Though not a mulch, erosion netting is used to hold mulch in place on steep slopes or in windy areas in order to get plants established there. The netting is usually a fabric similar to burlap with meshes a little larger, useful in anchoring mulch. Cheese cloth will also do a fairly good job. This netting protects against erosion and holds the soil in place until the seeds come up or other plants get established. Steel pins are used to pin the netting to the soil. Erosion netting is available through most good nurseries.

Organic Mulches

No other activity can do as much for the garden as the use of organic mulches. These are the best mulches and usually consist of straw, spoiled hay, grass clippings, old leaves or, in some cases, sawdust, paper, or old plant residues that have been run through a compost shredder.

Organic mulches function as a protective layer over the soil surface much like a thick carpet covers and protects a floor. But at the same time, they allow air and rainfall to pass through.

Organic mulches are biodegradable, and there are advantages and disadvantages to this. They are bulky and trouble to locate and get to your garden; they break down and have to be replaced continually. But the advantages far outweigh the disadvantages: because they are biodegradable, organic mulches furnish food for earthworms and microbes, both of which do many useful things for your soil. Organic mulch materials are often available free and so may cost less than other mulches. And, since most of us don't get nearly enough physical exercise, the labor of handling organic materials certainly doesn't hurt!

Organic mulches are a form of timed-release organic fertilizer. If you lift up a heavy hay mulch after it has been in place a few months in a produc-

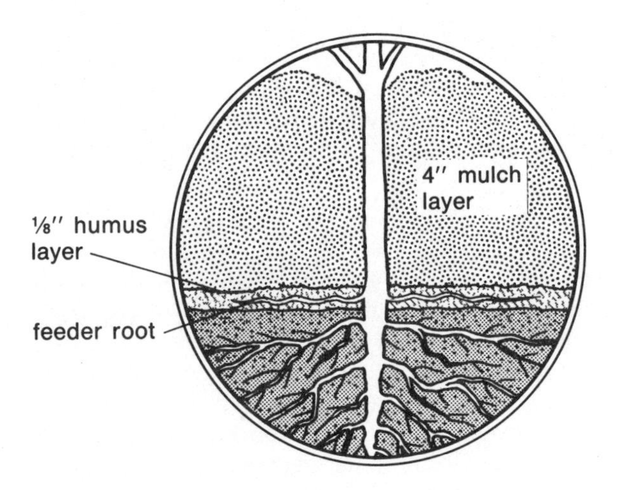

Figure 28. A magnified view of what happens to a plant when you supply a good mulch.

ing garden, you'll see two things: (1) a thick dark-colored layer of humus on the soil surface, and (2) plant feeder roots in this layer (Figure 28). Though you can't detect it, these feeder roots have thousands of root hairs feeding from the areas near the soil surface, and this is an ideal place to feed because under mulch there is *plenty* of three things roots need to perform best: oxygen, plant food, and moisture.

The thin layer of humus will be at the interface where mulch and soil surface meet. It is only a thin layer, but it's potent and is continually replenished from the mulch layer above. This is why you have to continue to add mulch on the surface from time to time. The original layer of mulch is slowly digested from the bottom by the life in the soil, and the height of the mulch layer gradually sinks.

Organic mulch—no tilling. Upon moving into a new house one spring I was dismayed to see instead of a garden spot an area with wild oats 2 feet tall. Having read about mulches and knowing a farmer with a big pile of spoiled hay, I decided to go for broke with mulch culture. I set up a rotating sprinkler on a stand in the center of the intended garden area, then spread loose hay 18 inches deep, watered it down well, and waited. The first sprinkling settled the loose hay down to about 9 inches. This 9-inch layer kept shrinking until in about 2 months it was only about 4 inches thick. When peeled up, you could see the humus layer, and earthworms scurrying for cover. When I raked my fingers across the surface, it was loose and soft, as if it had just been rototilled.

What next? You're ready to plant! The weeds have been killed by the mulch and the soil has been tilled and made soft by living things. But how can you plant in mulch? You can't. You have to windrow the mulch into neat rows to expose a band or slot of soil about 6 inches wide. This allows you to make a planting trench (Figure 29).

To do all this, drive a stake at each end of each row (make row centers 36 to 40 inches; 40 inches is probably better). Stretch a strong cord between stakes to mark the center line of the row. With a pitchfork, pull the mulch from under the string in two directions, piling up the mulch removed in the aisles to form the windrow. In the 6-inch wide strip where there is no mulch, make a planting trench, plant and cover seeds, then wait for the seedlings to emerge. When plants come up and have grown about 3 inches tall, you can pull the mulch back to the center line of the row to cover the bare soil (Figure 29, A).

A little later, when the young plants have grown up about 6 inches high, you can pull the mulch back level as it was originally (Figure 29, B). For some low-growing plants like lettuce there's no need to pull the mulch back level. Just pull it back so that no bare earth shows.

Weed Control

Keep bales of hay, straw, or whatever along the edges of the garden. Whenever you see a weed, don't go for your hoe, go for your pitchfork. Throw a forkful or two of hay on top of the weeds you see.

If your organic mulch material is not baled but is loose, place piles of the material along the edges of the garden. If the material is too fine or otherwise cannot be handled with a pitchfork, buy yourself a

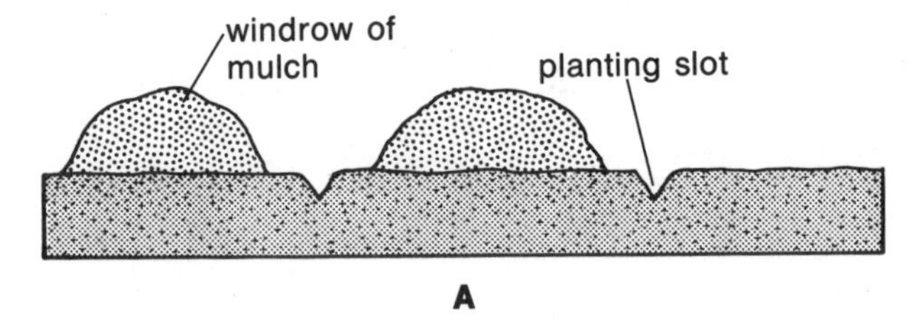

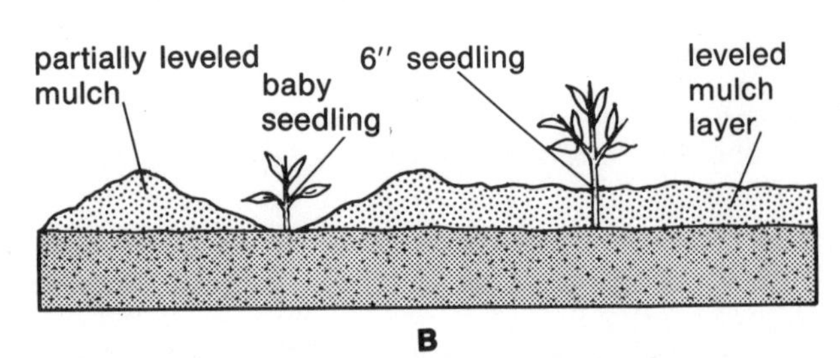

Figure 29. A—Windrowed piles of mulch separated to leave 6-inch slot for planting. B—Seedlings in mulched garden.

scoop shovel. This is a shovel with a huge scoop for moving light, fluffy materials.

When to Apply Mulch

In early spring, if the entire soil surface is covered, the soil will be insulated against warm solar radiation. If you want an early garden, this could be a disadvantage. In a fertile soil usually the thick layer of summer mulch will have thinned down so that by fall it is quite thin and by the next spring it is nearly gone. What is left by planting time in February or March can be windrowed to form a 12-inch wide strip of bare soil to soak up solar energy. Then when it really begins to warm up in May, more mulch can be added to give a thick layer between the rows.

You'll find that soil has "digestive power" for organic matter. When you first start mulching the soil, it "eats" mulch slowly. But after a year or two its digestive powers are remarkable. You'll have an army of earthworms and microbes to feed. Just remember that as you feed them they work for you.

Sowbugs and Mulch

What about sowbugs (pillbugs)? These interesting little animals (they're not insects) thrive under mulch and sometimes forget they're not supposed to chew on green, living plants (some textbooks say they eat dead organic matter only). I think it is a matter of coexistence. Sowbugs aren't supposed to attack living plants when plentiful supplies of organic matter are present, but I have seen them do it.

One solution for those unwilling to use any chemical control at all is to go ahead and plant thickly and hope for the best, then "take the lumps" if sowbugs attack. The bare 6-inch wide mulch-free strip should help somewhat since these little bugs must have a constantly moist environment.

The other solution is to use a spot treatment of the safest insecticide we have. This is carbaryl, sold under the tradename of Sevin, and for sowbug spot treatment you need "Sevin Insect Bait." Upon planting, scatter the Sevin insect bait thinly over the top of the planting trench. Even better is to wait until just before the plant sticks its head out of the ground, then apply the bait. The sowbug loves tender tips of newly germinated and emerged seedlings. If you put the poison there just before the tender growth of your plants appears, the bugs will get a treatment instead of a treat.

Mulching vs. Composting

Use of organic mulches has been called "sheet composting," but this is misleading. There are major differences between compost and mulch. One is temperature. In a compost bin the temperature rises dramatically and will get so hot you can't keep your hand in it. So-called "sheet compost" does not heat up. Earthworms and "sheet compost" can coexist; you'll never find an earthworm in a true compost pile—it's too hot.

Mulching more closely imitates nature than composting does. Nature does not build compost heaps; she drops all refuse in a thin layer on a more or less continual basis on the forest floor or as a sod on the prairie.

It's best to have both mulch and compost around. There's nothing like having some good, homemade, screened compost for making potting soil or for covering seeds in the planting trench.

Soil Compaction and How to Handle It

A fertile soil will grow healthy, productive plants. But to do this in an outdoor natural setting, the soil needs to be deep and well drained.

If you've ever put in a new lawn, you may have seen problems develop the following summer, problems like "bad spots"—those areas that begin to burn first before the next rain or irrigation or that may not respond to fertilizing.

These may be due to compaction of a soil layer 6 to 10 inches or even more under the surface (Figure 30). This layer is hard and tight. Roots penetrate it with difficulty, if at all. They tend to strike it and, not being able to penetrate it easily, turn sideways. A few roots may get through, but not enough to give a really good root system.

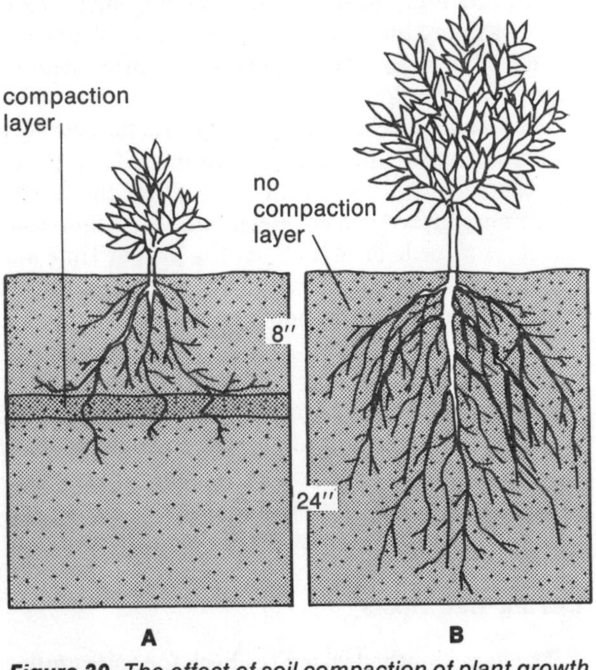

Figure 30. The effect of soil compaction of plant growth. A—Poor root system and stunted plant. B—Normal root system and healthy plant.

In a normal soil without a compaction layer, roots penetrate deep into the soil. Not only are roots deeper but there are more of them (Figure 30, B). In times of drought, the deep roots are still in moist soil and the plant survives.

Subsurface Drainage

During a heavy rain or a flood, the top 12 inches of a soil may become saturated. This means that all the pore space is filled with water and all air is crowded out. This condition is all right if it doesn't last too long. But if anything traps the water so it can't drain down through the soil, roots may suffocate.

A *compaction layer* interferes with downward movement of water, so it hurts subsurface drainage. The water may drain eventually, but it will take too long.

Hardpan

A more serious subsurface condition is the soil called a *hardpan* (Figure 31). This is a rock-like layer under the surface. It is so hard neither roots nor water can penetrate it. The top 12 inches of soil may drain well and the water move downward until it meets the hardpan layer, where it piles up because it cannot pass on through.

Claypan

In the humid areas of the South *claypan* is more common than hardpan. A claypan is a layer of tight clay under the surface. A typical situation is

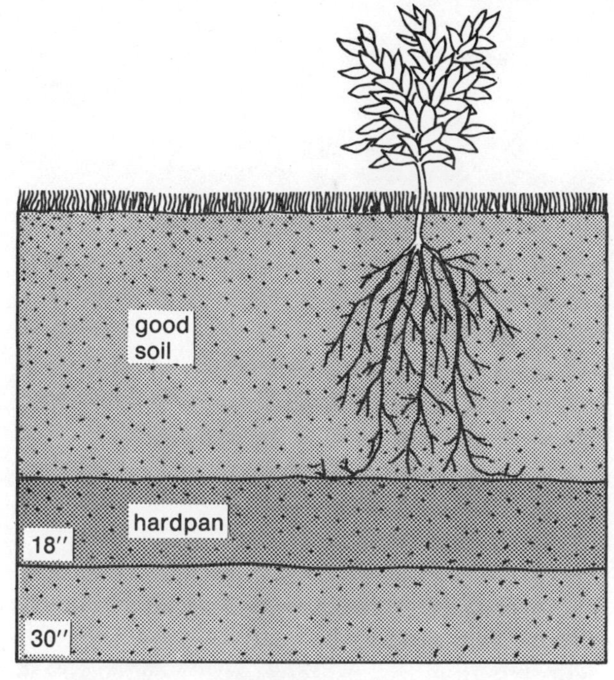

Figure 31. Hardpan, a tight layer of soil that prevents roots and water from penetrating.

a nice, porous, sandy loam sitting on top of a thick clay layer. This clay layer will admit roots, but only with some difficulty. It will admit water, but very slowly. And the same goes for air movement. This is why farmers smile when a dry spell is broken by a light rain falling steadily for 48 hours. But they frown when the same amount of rain falls in an hour or two. If they have a claypan soil, the slow, steady rain puts water deep into their soil. The intense rain, quickly over, ends up as runoff.

Testing for Soil Compaction

If plant growth is normal, you probably have little or no compaction. But if "hot spots" appear and areas in your lawn begin to "burn" during a drought while other parts of the lawn are normal and green, the soil may have compaction problems. During construction, heavy cement trucks, etc., may have driven over what is now lawn and compacted the soil.

There are two ways to test a soil for compaction; the soil auger and the water jet.

Using a soil auger. The soil auger is simply a king-size drill bit mounted on the end of a rod with a cross bar at the top for turning (Figure 32). It comes 2, 3, or 4 feet long.

To test soil compaction with a soil auger, pick a time several days after a good rain or irrigation, then bore a hole in the soil. You turn the auger handle until it "bites" the soil and moves downward about 3 inches. Pull it out, remove the soil, and continue to drill downward. Usually, the top 6 inches are easy; but at some deeper point the drilling may become more difficult. It suddenly "stiffens up." You have drilled into a tight layer of compacted soil. If you continue to drill, after a few inches or so the drilling will get easier. You will have drilled through the hard layer.

Using a water jet or probe. The water jet is a piece of pipe about 4 feet long with the bottom end pinched together to form a small opening, so when water comes out it will have lots of force. The top of the water jet has a handle and a hose fitting for hooking up to a water hose. When the water jet is held against the soil surface and water turned on full force, it will eat a hole downward into the soil (Figure 33).

As with the soil auger, in good soil the jet will move downward quickly and easily with little pushing from the top. But when a tight layer is struck, the going gets tough. You have to push down harder on the handle, and even so, the downward movement is slower.

Prevention of Soil Compaction

If you're putting a lawn or garden in a new yard, be sure the soil is moist (several days after a rain or after watering) and then rototill it 6 to 8 inches deep. This will unearth pieces of old sheet rock, etc., that may have been buried. It may also break up shallow compaction layers just under the surface.

Then apply a 2-inch layer of organic matter such as manure and rototill it in. After all this, grade the surface evenly to drain away from the house. Plant an adapted grass—Bermuda, St. Augustine, Bahia, centipede or carpet—at the proper time and manage it for good growth. This method will cost more to start out with, but it pays in the long run.

Stay off wet soil. Another way to lessen or prevent soil compaction is to stay off it when it's wet. Soil just after a heavy rain is soft and easy to leave tracks in. Every track you see means soil compaction. Stay off wet soil even if a lawn is growing. Use walkways for heavy traffic at all times.

Curing Bad Spots

The same methods used to check for soil compaction can also be used to cure it. The idea is to perforate the bad spots in a lawn with holes deep

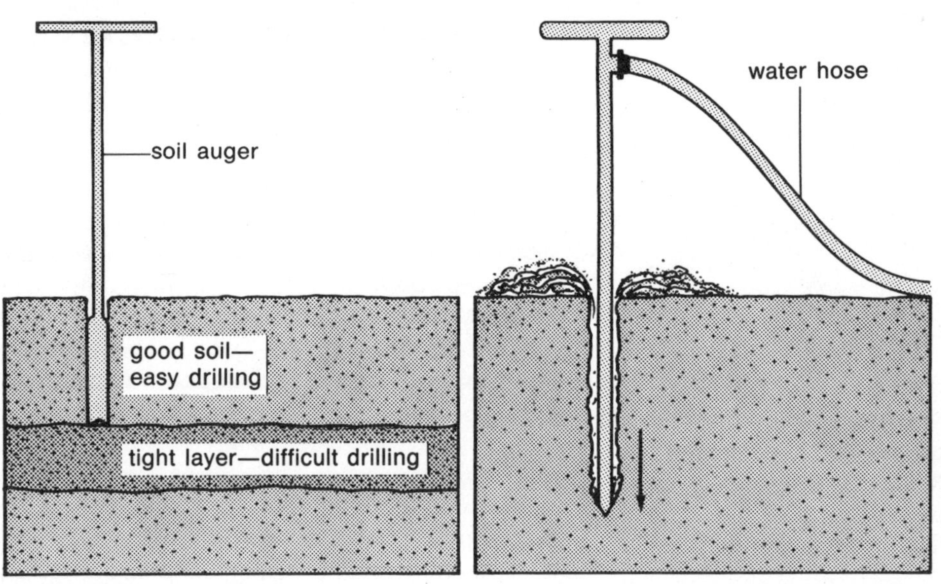

Figures 32 and 33. Using the soil auger to check for tight soil layers (left). A water jet can also be used to check for soil compaction or to drill a hole for fertilizer or for any other reason (right).

enough to get through hard layers. A 12-inch by 12-inch pattern will do a good job. A soil auger can be used but is enormous work. A water jet turns the job into child's play.

After putting holes down in your soil it is best not to leave them open. Fill them with a mixture of 1 part sand and 1 part compost or other organic matter.

I have treated sick lawn areas this way and had complete recovery in a single week. Sick, yellow areas greened up and disappeared.

If you want to use a soil auger, you can take some of the work out of it by motorizing the auger. Saw off the handle and buy an electric drill with a chuck large enough to accommodate the sawed-off top of your auger. Be sure it is a variable speed drill. As you run the drill at slow speed it will drill a hole for you. But you will still have the job of pulling auger, electric drill and all up out of the ground after about 3 inches of downward movement.

A factory-made water jet for drilling holes in tight soils.

Treating Large Areas

It is one thing to use a water jet to cure a small spot on the lawn. It's quite another to do several thousand square feet. On large problem areas where heavy tractor-drawn equipment can be used, a "subsoiler" is useful for breaking up tight soil layers. This is steel shank with a small point or chisel at the bottom that can be forced down into the soil and then dragged forward at this depth. It is best done when the soil is dry so tight layers will be shattered.

The Slot Method Prevents Compaction

In a backyard vegetable garden, this is a good preventive against soil compaction. With the *slot method,* you develop a growing zone 24 inches wide and 6 to 8 inches deep. You never have to walk on the fertile strip and pack it down. You do all your tillage, all your planting and harvesting, from the walkway. See pages 6-7 for details and illustrations.

Manure vs. "Green Manure"

For soil to be kept really fertile, it should receive additional organic matter at least once each year. This can be organic matter hauled in from the outside such as manure, spoiled hay, crop residues, or lawn clippings. Or it can be "green manure," a dense growth of plants grown just to be worked into the soil for soil improvement. This is called green manure because it is worked into the soil while still green, when the carbon content is still fairly low. Oats, sorghum, and cowpeas are sometimes grown for this purpose.

Green manure is for large areas where tractors can be used and for areas too big to cover with animal manure. For the average size home garden, it will be more practical to haul in enough animal manure for a ½- to 2-inch layer each year and work it into the soil.

There are three ways to work organic matter into the soil, (1) biologically (earthworms), (2) hand spading, and (3) rototilling. Of the latter two methods, except for very small spaces, the rototiller is ideal.

About Rototilling

Nothing is equal to the rototiller for working up soil to prepare a seedbed or working organic matter into the soil.

But here is where proper use ends and abuse begins. Once the back-breaking job of getting organic matter worked into the soil has been done, what is left for the rototiller to do? If your soil is really fertile, your crop will soon overtake weeds and shade them out with a little help from you and a hoe. And far less damage will be done to the soil than had you run a plow or rototiller through the aisles between the rows.

If you must use your rototiller to till the aisles between rows, set it just as shallow as possible and do it as few times as possible. Tillage over half an inch, and certainly over an inch deep, may cut feeder roots and will oxidize organic matter out of the soil.

The "No-Dig" Garden

If you must dig, do so as little as possible for less work, more results, and a more fertile soil. No-dig gardening is just about what it says. You only dig planting trenches or to pull out old stalks. Rototillers are out and so are spading forks. But earthworms are definitely in. Details are given on pages 58-60.

Power Equipment for Working the Soil

Gardeners with small yards don't need any power equipment at all, except a lawn mower and perhaps a small rototiller. But those with large yards and vegetable gardens will find power equipment most helpful in soil management and gardening.

About Rototillers

Rototilling is power-spading—it loosens and pulverizes the soil down to about 6 inches. When you spade by hand you thrust the tool down into the ground, pry a chunk of soil loose, lift it up and drop it so that it breaks up. Then you rake the area level. A good rototiller does all this in one pass if it is properly used.

Estate-Type Equipment

The Graveley tractor and Mainline rototiller are top contenders here because they are powerful, versatile, have automotive-type transmissions and other good engineering features. Ariens is also in this class.

These mini-tractors will do a good job of rototilling, but they can do much more—mowing, cart pulling, plowing, and spraying. They're expensive but worth it if you have large and varied jobs. They employ advanced engineering features such as enclosed chain or gear drives (no belts).

Large Gardens

If you have a large vegetable garden or lawn, and rototilling is the main chore, the Graveley or the Mainline mini-tractors could still be used to good advantage if you don't buy all the attachments.

Other rear-tine tillers of 6 horsepower or more are also suitable.

Small Gardens

Any tiller (except the Graveley, which is too large) could be used, but it is difficult to justify much more than 3 horsepower or $300 for a power tiller.

Table 19
Comparison of Rototiller Performance*

Factor Evaluated	Brands/Power/Performance†				
	Honda (3 hp)	MF (5 hp)	Mainline (8 hp)	Rotohoe (6 hp)	Autohoe** (4 hp)
Ease of handling and turning	6.5	6.5	5	4	2
Leaves no tracks	7	2	8	7	2
Ease of gear shifting	6	7	7	5	—
Change tillage depth	7	7	6	5	5
Change handle position	7	6	7	2	—
Average Score	6.7	5.7	6.6	4.6	3

*Testing and evaluation made by an impartial committee from a local garden club at Big Spring, Texas, April 1976. MF, Massey-Ferguson.

†10 = perfect score

**Autohoe has hoe-like blades instead of tines. Vegetation inter-twined making operations as a tiller impossible in this test.

Rear-Mounted vs. Front-Mounted Tines

Tillers with rear-mounted tines are preferable (with one exception) since they do not shake you up the way front-tine machines do and they leave no wheel tracks.

The exception is the Honda, with middle-mounted tines. This tiller has its tines sitting directly under the engine, and no rear wheels to leave tracks. It has a rear tripod-like shank that drags behind to stabilize the tiller while operating.

Rototiller Performance Testing

Don't buy any tiller until you have field-tested it. In 1976 an actual field performance test was made at Big Spring, Texas. Of the 12 manufacturers invited to make available one of their tillers for this test, five responded. These tillers were lined up in the same field and each was used to till approximately 200 feet of row. An impartial committee from a local garden club evaluated performance. Results are shown in Table 19.

The Graveley was not included in the performance testing since the basic unit has a rotary plow and not a rototiller.

Cost and performance information on ten brands of rototillers are given in Table 20.

If you can't afford a high-quality tiller, don't buy. In the long run it's worth extra expense for better performance.

Should You Buy a Rototiller?

I have worked a 1,000-square foot garden using only a spading fork. As I broke my back spading the garden I dreamt of power equipment to take the load off my shoulders. Once I got a rototiller, I found I could do the whole job in an hour or less. Then I parked the tiller until next season, when it would be time again to prepare a seedbed. You should think about this.

Contemplate your $400 to $800 investment sitting there gathering dust month after month, not working. You'll begin to wonder. If you have a garden 2,000 square feet or less (660 linear feet of row), try to rent or borrow a tiller once or twice a year to prepare a seedbed using the slot method. A good tiller may cost you $600. On the other hand, for perhaps $20 a year you can buy the use of a tiller when you need one.

Remember that it is not necessary to rototill the entire surface. A rototilled slot 18 to 24 inches wide along the row center will be enough.

If you do elect to buy a rototiller, it's probably best to go through a dealer and let them hassle with adjustments, parts, etc.

Engineering Features to Look for in Large Tillers

1. 4 to 8 horsepower engine.
2. Transmission: automotive-type, gear shift, with reverse and two forward speeds (low and high). Additional speeds don't hurt, but you don't really need them. You need an easy-to-shift reverse to back out of tight places and reverse the tines to unwind vines or stalks that get wound up around the rotor shaft.

Table 20
Cost Comparison of Rototillers*

Brand	Tine Location	HP	Price	Performance Rating†
Honda	Middle	3	$ 390	6.7
Massey-Ferguson	Front	5	300	5.7
Mainline	Rear	8	800	6.6
Rotohoe	Rear or Front	6	400	4.6
Autohoe	Front	4	300	3.0
Ariens	Rear	7	No data	Not tested
Graveley	Rear	7.6	1100	Not tested
Herter	Rear	6	500	Not tested
Mighty Mac	Rear	7	620	Not tested
Troy-Bilt	Rear	6	600	Not tested
Yellow Bird	Rear	3	370	Not tested

*Prices based on 1976.
†From Table 19

Before you buy the tiller, insist on operating it to see if the gears can be shifted easily. A hard-to-shift gear on a rototiller is an abomination and leads to swearing.

3. Tines at the rear. Rear-tine tillers are much smoother to operate (less vibration). They also leave no wheel tracks nor an uneven surface that has to be raked smooth. A good rear-tine tiller leaves a level, smooth surface to plant (don't make the mistake of planting a newly rototilled bed unless you firm the soil over the seeds you have planted).
4. Tiller should be purchasable fully assembled and ready to operate or, if by mail order, one that is nearly completely assembled and takes less than 30 minutes to put together.

   Check this point very carefully and get in writing from the dealer just what is necessary to assemble the unit and how long it takes.

   It is better to buy from a dealer where you can pick up the tiller fully assembled, gassed and oiled, and run it before you sign the check, just the way you buy a new car. The only safe alternative is to talk to a nearby gardener who has ordered the same tiller unassembled, put it together, and is willing to come help put yours together.
5. Look for versatility if you have lots of different kinds of jobs to do. Lots of mowing? Need a snowplow in winter? Need a PTO (power-take-off) pulley to drive belts that will pump water or grind grain, etc.? Need a pull cart to haul material around the place? If so, look for the tiller that has lots of attachments.
6. External ignition points. Do you like to do your own simple repairs and tune ups? Have you ever tried to pull the flywheel off the top of a lawn mower engine to get at the points? It's no fun at all.

   Only the larger, more expensive tillers have external automotive-type ignition points. This saves lots of time and trouble in keeping the engine tuned and running at peak efficiency.
7. Consider a 4-wheel tractor for the small farm or large estate of 1 to 5 acres. Rear-mounted rototiller attachments and a rotary mower attachment are available. Some small tractors will drive a small flail-type mower that facilitates very close mowing of tough grass.

Limitations of a Rototiller

A well-designed rototiller will do a fine job of digging up and pulverizing your soil from a depth of 2 to about 6 inches. It will mix organic matter into the soil. It may be able to do hilling and make furrows.

But it won't do all your gardening chores. It will not make compost, nor will it serve as a compost grinder. To make compost, you must make a pile, usually in a compost bin. To grind compost, you will need a separate machine called a compost grinder.

Compost shredders or grinders

It is not advisable to buy a combination rototiller/compost grinder. The conversion from tiller to grinder is too difficult and time-consuming. Get a separate machine instead.

A compost grinder reduces coarse pieces of organic matter, such as cornstalks, into small, neat piles of small pieces. Compost grinders also do a good job of shredding old newspaper for mulching or for use in earthworm beds.

Old, dry, hard chunks of manure, if run through a compost grinder, work much better in a compost heap. Compost-making is speeded up if residues are first ground.

My Choice of Rototillers

Based on wide observation and performance testing, if I were to buy a rototiller today I would look for performance, engineering, and price, and the choices would be as shown below.

1. Small garden (100 square feet or less): do it by hand.
2. Small- to medium-size garden (100 to 1,000 square feet): rent or borrow a rototiller; if that's not possible, buy a Honda.
3. Medium- to large-size garden (1,000 to 2,000 square feet): Honda.
4. Large-size garden (2,000 to 4,000 square feet or more): Mainline with attachments.

The above choices are based on present data and could change with further observation and testing.

Tips on Watering

The Root Zone

The soil where the roots of plants are growing is called the *root zone.* The soil in the root zone is somewhat like the gas tank of your car: it stores water for future plant use.

In times of drought you fill up the root zone by irrigation. When the soil has been filled with all the water it can hold, it is said to be at *field capacity*. The plants use up the water like your car does gas, and after a while the "tank" is only half full; half the stored water has been used. At this point, water should be provided for best plant growth. If you live in the East, you hope it will rain; in the West, you turn the sprinklers on.

Soil Moisture Gauges

The root zone does not come equipped with a gauge like an automobile to tell you how much water is left in the "tank." But you can buy and install a gauge if you like. It is a long tube fitted with a dial which tells you at a glance the status of soil moisture. These devices are called *tensionometers*.

The plants themselves are fairly good gauges if you learn to "read" them. In response to low moisture conditions in the soil the leaves of many plants change color to a darker green. Other plants, such as corn, respond by rolling up of the leaves when they get thirsty. A lawn responds by a very light "burning" of the leaf tips.

How Deep Does an Inch of Rain Wet the Soil?

A little water goes down a long way in a sandy soil. A 1-inch rain will wet a sandy loam soil down about 16 inches. But the same 1-inch rain, if it all goes into the soil, wets a clay soil down only about 5 inches (Figure 34).

Basic Soil-and-Water Relations

Let's say you have a planter box 12 inches square by 12 inches deep with a plugged drain hole at the bottom. Fill the box with dry loam soil and add water until it barely rises to the soil surface.

Then pull the plug. Lots of water will drain out, but about 2½ gallons will remain in the soil. This stored water would be equal to a layer about 4 inches deep if it were stacked on the 1 square foot of surface at the top of the planter box (Figure 35).

The soil is now filled up to its full capacity, its *field capacity*. It is like a gas tank you have just filled up to the "full" mark. Plant some live plants in the planter box, and under good growing conditions, they'll begin to withdraw the stored water. After a while, unless it rains or your irrigate, plant growth slows down. Still later, if no water has been applied, the plants wilt and die. This point is called the *wilting point*.

If you test the soil at wilting point, you'll find part of the original stored water still in the soil (about a gallon, or 2 inches). It's there but "locked up" so the plants can't get it. Only about half the

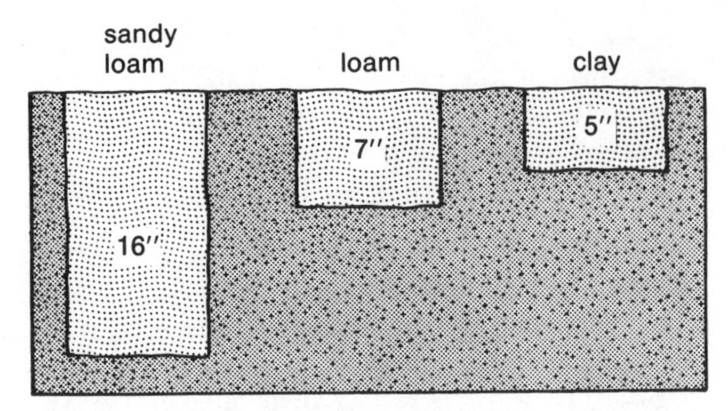

***Figure 34.** Depths to which 1 inch of rain will penetrate clay, loam, and sandy loam soils.*

original amount stored is available to plants. We call this water that is available to plants the *available water capacity* or just *available water*. For this soil, the available water was about 2 inches (about a gallon of water).

Effect of Soil Moisture on Plant Growth

Just after a good rain or an irrigation followed by drought, the growing plant goes through the following stages:

Stage 1: Readily Available Water

The soil in root zone starts at field capacity. Total available water in a typical silt loam soil equals 2 inches per cubic foot. Plant roots begin to extract water stored in the soil. With a typical silt loam soil (like in our planter box), a layer of water about 1 inch deep (2½ quarts) is readily available for plant use per square foot of area 1 foot deep (Figure 35).

Stage 2: Slowly Available Water

Ideal time for rain or irrigation to replace water used by the plant (about 1 inch). All the readily available water has been used up (about 50 percent of the total available water). The roots begin to work a little harder to get the remaining stored water. Plant growth may begin to slow down.

At the end of this stage, 75 to 85 percent of the total available water has been used up and plants are stressing for water. Plant leaves may turn a dark green or bluish-green; leaves may roll up.

The end of this stage is an S.O.S. to the gardener. You have only a day or two left. Pray for rain or get ready to turn the sprinklers on. Cloudy weather and cooler weather may help some.

Stage 3: Temporary Wilting Point

Nearly all of available water is used up. It's extremely difficult for plant roots to extract water

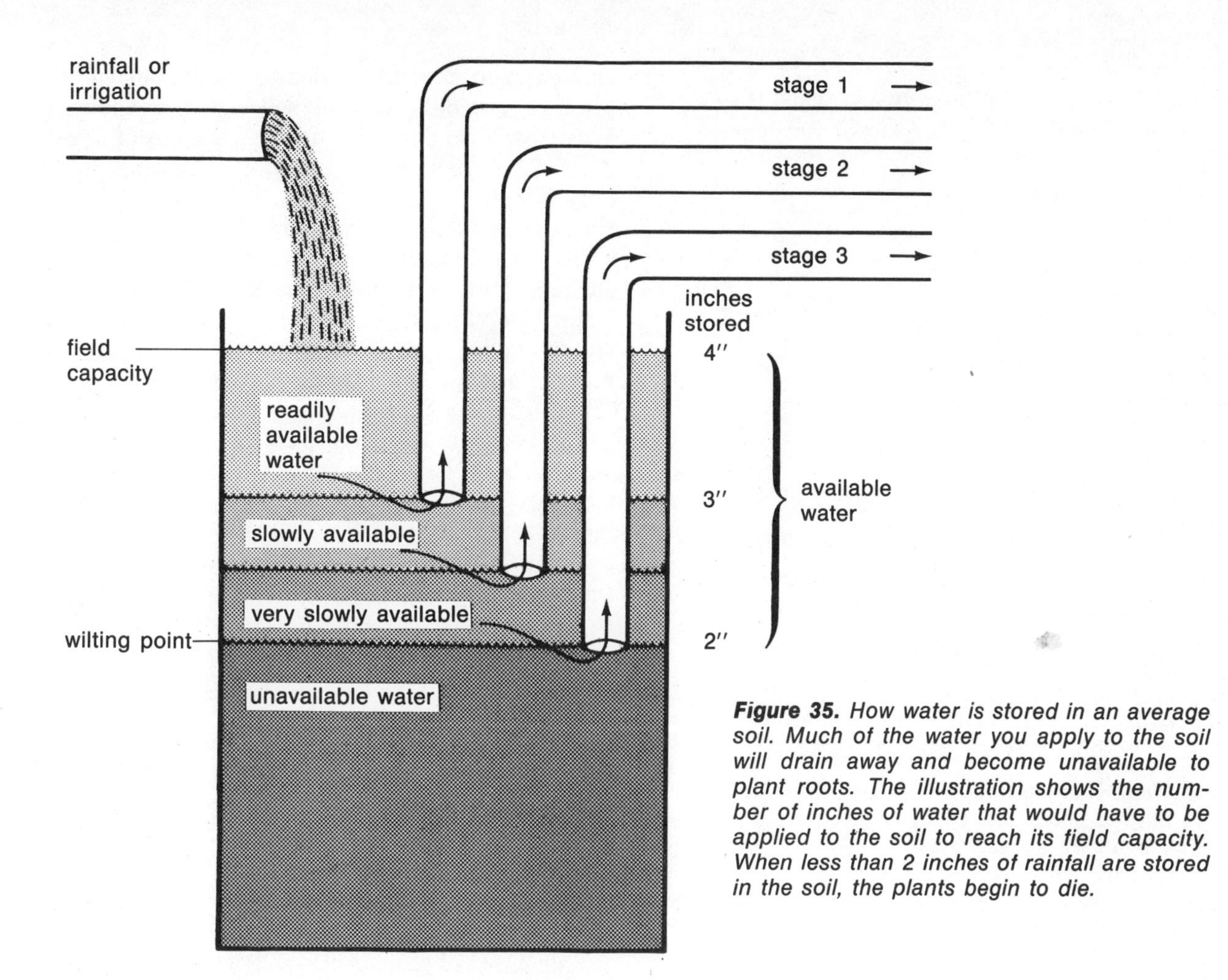

Figure 35. How water is stored in an average soil. Much of the water you apply to the soil will drain away and become unavailable to plant roots. The illustration shows the number of inches of water that would have to be applied to the soil to reach its field capacity. When less than 2 inches of rainfall are stored in the soil, the plants begin to die.

from soil. Growth stops. The plant is hanging on to a slim thread of life. Desperate situation. About to die from thirst. Rain or irrigation is needed immediately. Two inches of rain per foot of rooting depth are needed to fill soil back up to capacity. For a day or two, plant leaves may wilt during day and recover at night or on a cloudy day.

Stage 4: Death by Thirst
All available water is gone. Plants wilt permanently and will not recover. Remains go to the compost heap.

When to Irrigate

An important rule of thumb is to irrigate when about *one-half* the available water stored in the root zone of the plant has been used. How do you know when? Use the "feel test." To do this, you need a spade, soil tube, or soil auger to take a sample from a depth of 8 to 12 inches (or 4 inches for shallow-rooted crops like onions). Take a handful, squeeze it firmly, and open your hand. If it forms a ball easily, you have plenty of moisture; if it balls only with difficulty, it is time to irrigate.

How Often to Irrigate?

A shallow rooting zone, as with onions, means you'll have to irrigate much more often than with a deep-rooting crop like melons, which may root 4 feet deep. The more shallow the soil or the more shallow the rooting habit of the plant, the less water is stored and the more often irrigation is necessary.

Soil texture is a big factor affecting necessary frequency of irrigation. If you have tried to maintain a good lawn on a sandy soil, you know what I mean. You can only store about ½ inch of available water in the top foot of a sandy soil, and irrigation is needed when half of this has been used. If the weather is hot, dry, and windy, and the days are long and sunny, your lawn will run out of water in a day or two. The more fortunate gardener with a loam soil can go four times as long—about a week—under the same weather before having to irrigate or have his lawn burn up (Table 22).

High humidity curtails plants' need for water; so do still, cloudy days, Short day length, as in spring and fall, also slows down their water use.

Table 21
Feeder Root Zones of Some Plants When Roots are not Restricted by Claypans or Hardpans

Very Shallow (6 to 12″)	Shallow (To 18″)	Medium (To 48″)	Deep (To 60″)
lawn grass	beets	shrubs	trees
radishes	carrots	berries	grapes
onions	corn	beans	asparagus
chives	lettuce	cucumbers	melons
	potatoes	peppers	pumpkins
	turnips	peas	squash
			tomatoes

Table 22
Available Water-Holding Capacity of Soil

Soil Texture	Inches of Available Water Stored per Foot of Depth	Inches to Apply When One-Half of Available Water Has Been Used*
Sand	0.5	¼
Sandy loam	1.0	½
Loam	1.5	¾
Silt loam	2.0	1

*These figures are inches per foot of root depth. Plants with deep taproots would require more water; plants with shallow roots, less.

Plants use the most water in June, July, and August. They may use 1/10 to 4/10 inch per day, depending on the weather. For precise information of daily water use by plants in your area, call the nearest office of the USDA Soil Conservation Service. In the South's humid areas in mid-summer the daily water use rate may be about ¼ inch per day (an inch will last about 4 days), or about 2 inches per week.

How Much Water to Apply

Apply just enough water to refill the root zone of the plant to field capacity (Tables 21 and 22). If your irrigation water is not relatively salt-free (under 500 ppm of salts or if high-sodium water), apply 20 percent extra water for leaching.

This applies to pot plants as well. There is no way to keep the potting soil of a houseplant "slightly moist." Always water a pot plant until a little water runs out the drain hole at the bottom. If you don't, the soil will "salt out" and your plant will suffer salt burn and perhaps die.

How to Check Up on Your Irrigation Job

The use of tables to figure out how much irrigation water to apply are good, but they're not enough. You need to be able to check your soil to see how you did. Was the soil wet deep enough? Was it wet too deep? No use to wet the soil 2 feet deep if the plant you're growing is rooting down only 1 foot.

One way to check your irrigation is to set two coffee cans about 3 and 8 feet away from the sprinkler; after 15 minutes check the depth of water caught in the cans. This will be the depth of water applied over the entire area, provided you moved your sprinkler from set to set properly (see "How to Use a Sprinkler").

The best way to check on your irrigation is to "ask the soil" how deep the water penetrated. The *moisture probe* is the tool for this job. The moisture probe is a long, slender steel rod with a cross arm on top and a slightly sharpened bottom tip for thrusting downward into moist soil. As long as the soil is moist the probe can be pushed down easily. But the moment it strikes dry soil the probe reacts like you had struck concrete.

You really need two probes, one for shallow-rooting crops like a lawn and another for deep-rooting crops like shade or fruit trees. For lawns, a screwdriver with a blade 8 to 10 inches long works fine. For other crops, get a steel rod about 36 to 48 inches long.

A lawn should be irrigated until you can push the screwdriver probe down about 8 inches. A tree

A soil moisture probe for checking the depth of water penetration.

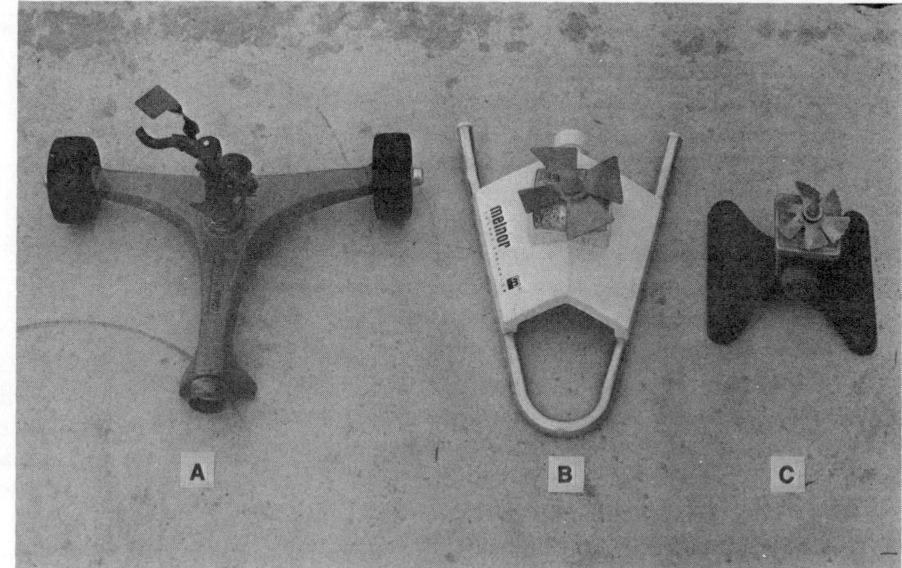

Three good sprinkler types for home gardening. A—Revolving head, pulsating sprinkler (flap is for adjusting wetted diameter, pins adjust to wet an area from ¼ to a full circle) useful for large areas. B and C—Revolving head, non-pulsating sprinklers for small areas.

should be irrigated until you can push the long probe down 3 to 4 feet, if you have that much soil depth. Be sure deep watering extends a little beyond the drip line.

Sprinklers

Sprinklers are best anywhere the water is plentiful and of good quality and slopes are gentle, but be sure you get a good sprinkler. The best all-around sprinkler is a pulsating, revolving-head type mounted on wheels and good for large areas. You also need a smaller, revolving-head, non-pulsating type sprinkler, preferably a sled-type on runners for easy moving.

Avoid the oscillating, back-and-forth type sprinkler. In fact, if you have one, try to leave it in your driveway to get run over so you have to go out and replace it with a really good sprinkler. The oscillating type sprinkler throws water too high and puts it out unevenly. In areas with poor water quality I have seen leaves burned off shade trees where this kind of sprinkler threw water. In sprinkling it is best to keep the water output close to the ground and in fairly large drops.

The pulsating type sprinklers come with a flap that enables you to control how far it throws the water. They also come with an adjustable clip and pins that allow you to wet an entire circle or any part of a circle such as a half circle or wedge-shaped area.

The small revolving-head, non-pulsating type sprinklers will wet an area as small as 24 inches square. These are handy for small, crowded home garden areas.

Other Watering Equipment

The water jet puts a hole down 6 to 48 inches deep far more easily than a crowbar. And it doesn't injure so many roots. See page 66.

Coiling and uncoiling a garden hose can become a chore unless you have a way to reel it up. The hose caddy—a revolving reel with a handle to turn so you can wind up your hose neatly and quickly—

A "hose caddy" for coiling and storing your water hose. This is especially useful if you use long, flexible drip irrigation lines.

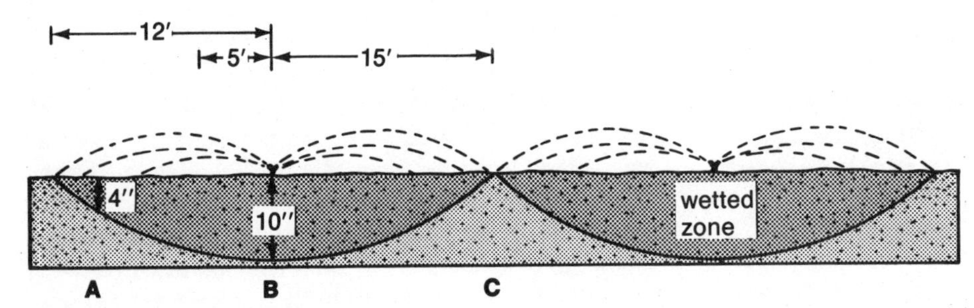

Figure 36. The wrong way to use sprinklers—by not overlapping the sprinklers, water is applied unevenly. Zone A will only be wet to a depth of 4 inches, B to 10 inches, and C is moistened only at the soil surface. (Not to scale.)

is the perfect item. It can be mounted on wheels for portability. It is especially useful for winding up and storing a drip irrigation line.

How to Use a Sprinkler

Don't try to water your lawn by holding a hose in your hand. It is emotionally satisfying but not very effective. No one can stand there long enough to apply sufficient water and do it evenly. Let a sprinkler do the job for you; but you must use the sprinkler properly for best results.

If you use a sprinkler properly, you can apply a uniform depth of water over a large area. If you don't, you can water some spots 10 inches deep and others only 4 inches deep (Figure 36). This is due to an unfortunate characteristic of any sprinkler: the circle wetted does not all get the same amount of water. The edges of the circle hardly get any.

What to do? Irrigation engineers have found that "sets" of sprinklers must be overlapped 50 percent for uniform application of water (Figure 37).

First, let's define a term. When using a sprinkler, we set down on a point and let it run awhile. This is called a *set*—a sprinkler sitting on a given point and running for a given length of time is doing a set.

What is a 50 percent overlap? Suppose set No. 1 (Figure 37) throws water 10 feet (a 10-foot radius).

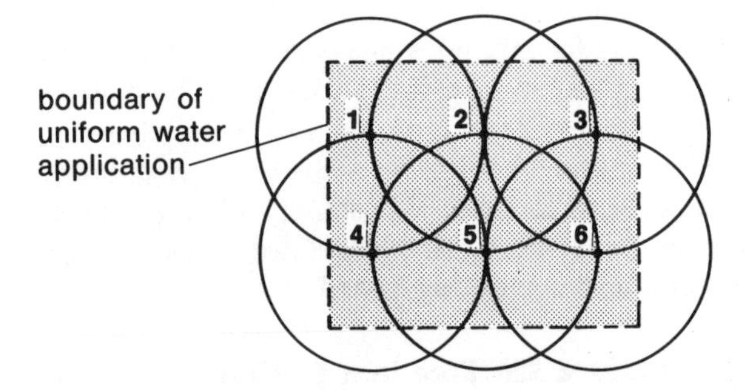

Figure 37. An overhead view of the proper positioning of the sprinkler(s). This will give a 50 percent overlap of sprinkling, which will provide a uniform application of water. Numbers indicate each position.

When you move the sprinkler to set No. 2, move it 10 feet, to the edge of where water landed from set No. 1 (Figure 37). It looks as if you are wasting water on areas already watered, but it is an illusion. If you water "edge of wetted circle to edge of wetted circle," you'll end up with a mighty sick and dry lawn or garden at point C, Figure 36. Make set No. 1 at X and set No. 2 at C.

I have been called out on sick lawn problems to find the trouble due to failure of the irrigation contractor to install the sprinklers with a 50 percent overlap. If a sprinkler throws water 15 feet (a 15-foot radius), all sprinklers should be set on a 15- by 15-foot pattern.

Watering Trees

The same principle applies when watering trees by sprinkling. Remember, only 5 percent of the water used by a tree is in the trunk area. Put most of your water where it is needed most—out by the drip line—and overlap sprinklers 50 percent.

Drip Irrigation

In drip irrigation water is applied drop by drop over a long period of time to wet a circle that is seen to be moist but not wet. When you water a flower bed with a hose, you flood the area and the water is pulled down into the soil by gravity. If it is sand and you lay the hose down to let it run, most of the water goes straight down and spreads out little.

Drip irrigation is different. It spreads out in all soils, and rather quickly in sand. Set a drip irrigation outlet on a spot of sandy soil, let it run 4 hours, and you have a big, moist circle perhaps 4 feet in diameter. No water can be seen, just moist soil. Water from drip irrigation moves outward from the source by a natural force called *capillary action,* and not by gravity. This means you can install a drip irrigation outlet a foot under the ground and the water will move outward in all directions from the source, including upward, against the force of gravity.

The simplest, least expensive drip system is simply a garden hose 50 to 100 feet long with emitters punched in every 2 to 3 feet. The *emitter* is a small, dart-like device that is inserted into a small hole in a hose to allow the water to drip out slowly and evenly. Several companies handle drip supplies for the home gardener, including Submatic, Inc., of Lubbock, Tx.

Where to Use Drip Irrigation

Drip irrigation was designed for desert agriculture. It was first developed in Israel, moved into California and western Texas, and now is used widely in such places as Florida and other eastern locations.

Drip irrigation is especially useful where (1) water is low-quality (2) irrigation water is from deep wells, is scarce, and costly to pump, and (3) special problem areas, such as sand dunes, exist.

Moisture Meters for Houseplants

It is always a problem to tell just when a houseplant needs watering. One way is to simply stick your finger down into the soil to the first or second knuckle to feel how moist or dry the potting soil is. The trouble with this method is that it is highly subjective. What is moist to one person may not be to another. And who knows how the plants feel about it.

I prefer to use a more objective method, such as a moisture-sensing device. During the past several years several brands of electronic moisture meters have come on the market for checking moisture content of soils, especially potting soils for container plants.

These meters have a long, slender probe about 8 inches long. Near the tip of the probe is a plastic band insulator. This bottom half-inch is the sensor that detects the moisture content. When inserted into the soil, moisture allows electric current to flow through the sensor, which is hooked up to a needle and a calibrated dial. The more soil moisture present, the higher the reading.

The manufacturers include with the meter a chart relating the readings of the meter to a reading most desirable for various houseplants. You check your plants about once a week to see which ones need watering.

The only bad thing about these meters is that they can give a false reading due to soil salinity. Sometimes you can get a reading indicating high soil moisture, but when you check with your finger, you find the soil quite dry. You can avoid this by proper watering to avoid salt build-up (see pages 11-13).

When selecting a moisture meter, look for one with the probe and dial separated, not one with the probe rigidly mounted to the dial and the dial sitting at the top of the probe. It is easier to use the meter where the probe is connected to the dial with a wire; this way you can reach up, insert the probe in a hanging basket but hold the dial at convenient eye level to read. Also look for a large, easy-to-read dial with the divisions and numbers clearly visible. Among the better meters is the "Instamatic Moisture Meter" put out by the CASCO Co.

During two years of testing I have found the moisture meter quite satisfactory for telling when to water houseplants. It is mighty handy to be able to push the slender probe down 6 to 8 inches deep into the root zone of a giant rubber plant in a 12-inch pot and find out what the soil moisture conditions are at that point.

The Bottom Line: Putting These Words to Work

Knowing soil improvement principles is one thing; applying what you know is another. And applying these principles is of utmost importance.

We have seen how to improve soil structure, how to load the soil with food, how plants get the food and use it. Now let's look at applying these principles to a specific problem at home: properly planting a tree or shrub.

Assemble Materials Before Planting

Before planting a tree or shrub, get your materials together. You'll need a shovel, preferably a round-point shovel for digging and a square-point shovel for mixing; a heavy-duty wheelbarrow for mixing soil; a tarp if planting is to be done in sod; a 20-gallon container, such as a garbage can, for mixing starter solution; some high-phosphorus fertilizer such as superphosphate; soil amendment such as peat moss, fine shredded bark or vermiculite; and a planting board.

Don't Plant a $20 Tree in a $2 Hole

A good tree or shrub doesn't come cheap. It deserves a good spot to grow in. Proper treatment of the planting hole pays big dividends.

First, dig the hole 1½ to 2 times the size of the root system if it's a bare-root tree or shrub (Figure 38). Why so large? So you can modify and improve the rooting zone soil for the young plant to give it the ideal conditions that will get it off to a good start. An exception to this is the pecan tree; when transplanting a pecan, you dig the hole 36 to 40 inches deep and prune the root system to fit the hole.

If the tree or shrub to be planted is a B&B (ball-and-burlap) or a container plant, measure the diameter of the root ball and dig the hole 1½ to 2 times larger; for example, if the root ball is 12 inches in diameter, dig the hole 18 to 24 inches wide and deep.

Remember, always plant the tree or shrub at the *same depth* at which it grew in the nursery. By looking at the plant you can see where stem and root merge. This is the "soil line." On many plants the stem will make a crook 6 to 12 inches above this soil line. This crook is where the plant was grafted or budded. Certain plants—roses, for example—are planted deeper than their original soil line before transplanting, but this is only done in the more northern areas of the South.

Planting Depth

To be sure you set the transplant at the proper depth, use a notched planting board. Get a piece of straight board about 4 feet long. A 2x4 is good. Halfway between the ends of the board, cut a V-notch. Place the notched planting board over the center of the planting hole to ensure planting at the exact depth the tree grew in the nursery.

Once the planting board is across the planting hole, with the V-notch in the center, have a helper place the stem of the plant in the V-notch and adjust the tree or shrub upward or downward so its soil line point is against the bottom of the planting board. Fill in around the trunk while the helper holds the plant in place. When the soil is all filled in around the plant and the planting board is removed, the plant will be at just the right depth.

Soil Mix for the Bottom of the Hole

You can plant the tree or shrub in the soil that you dig out of the hole and often have pretty good results. Better results can be had by mixing the excavated soil half-and-half with peat moss or other soil amendment. But the *best* way is to use the half-and-half mixture for the "root zone" and a mix of ¼ soil amendment and ¾ soil for the bottom of the hole. This is "transition zone" soil and helps roots make the transition from the rich, ideal conditions of the root ball to the non-ideal conditions of the native soil.

If you're planting in a lawn area, it's best to remove the sod from a 4-foot circle. Baby trees and shrubs do best without grass roots competing.

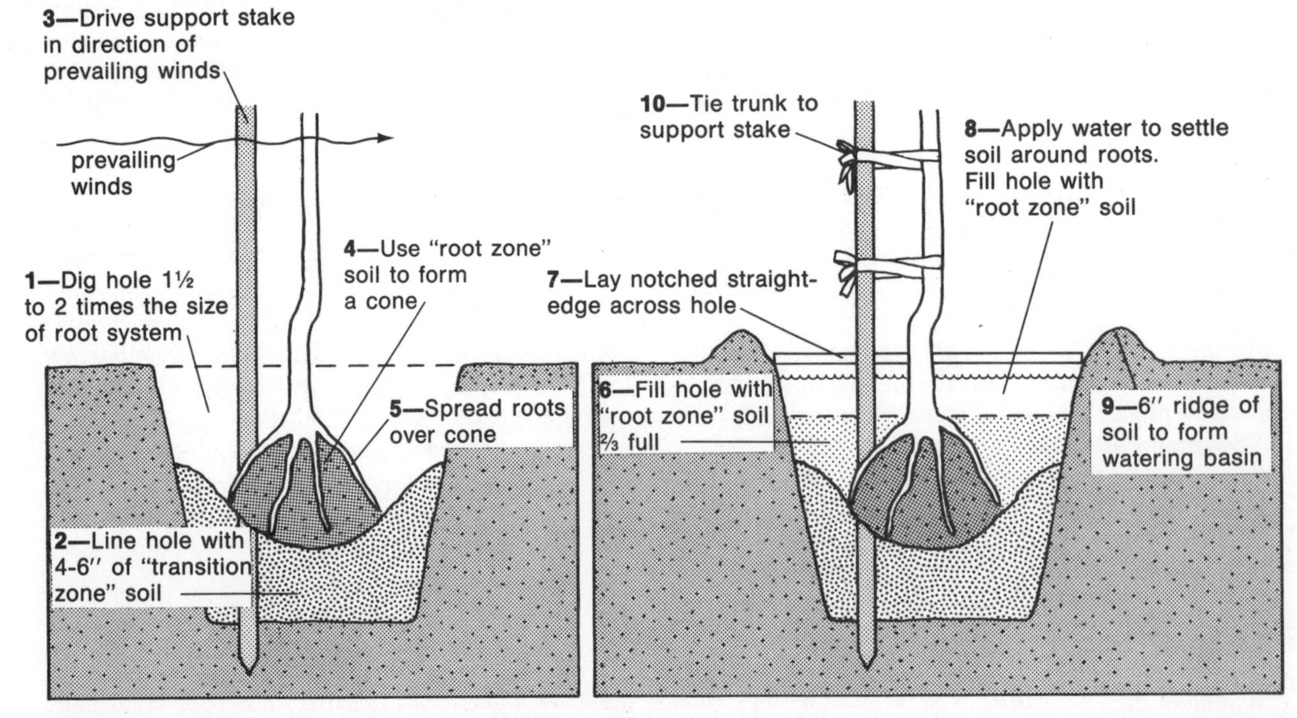

Figure 38. Ten steps to planting bare-root trees and shrubs.

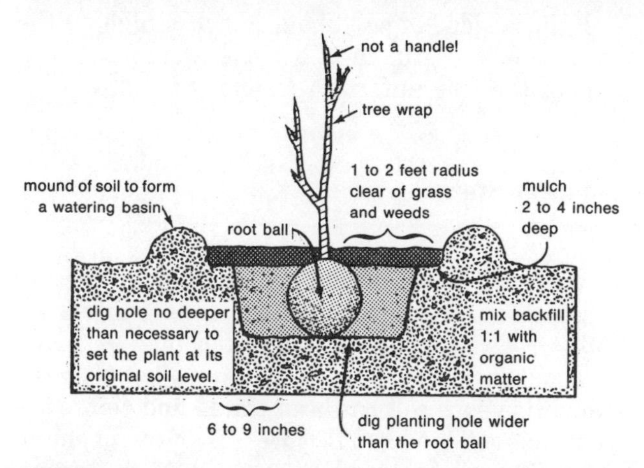

Planting in sandy, well-drained soils or in areas of low rainfall.

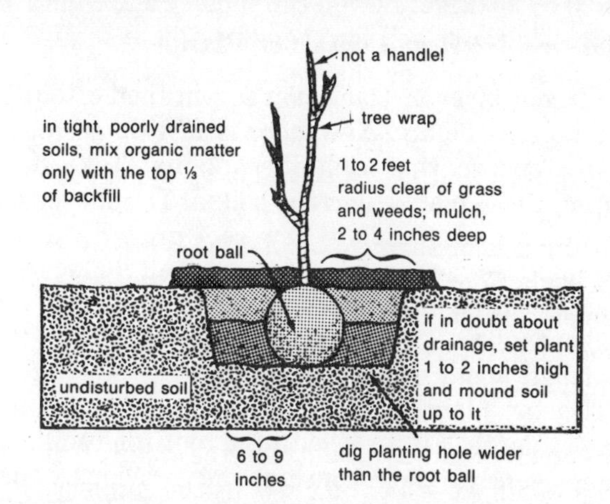

Planting in clay or poorly-drained soils in areas of high rainfall.

Dig the planting hole and place the dug out soil on a tarp. Move the wheelbarrow up, along with the soil amendment, and mix enough of the 1 to 3 mix (transition zone soil) to line the bottom of the hole 6 to 8 inches deep. Mix it well, then mix again adding water to thoroughly moisten.

Add the transition zone soil to the bottom of the hole. Then mix up the half-and-half root zone soil and moisten. Make a cone of soil so the roots can be spread out, and have the plant rest on top of the cone (Figure 38). Firm the cone well to get rid of air pockets.

With the plant held at proper depth, fill in ⅔ with the half-and-half root zone soil mix. Tamp gently with a stick to force out air pockets. Then fill hole with starter solution (see page 55). This will settle the soil and force out air bubbles. Let go of the plant stem for a few seconds to see if it remains at the proper depth. It may settle. If so, use more soil and work in under the roots while pulling

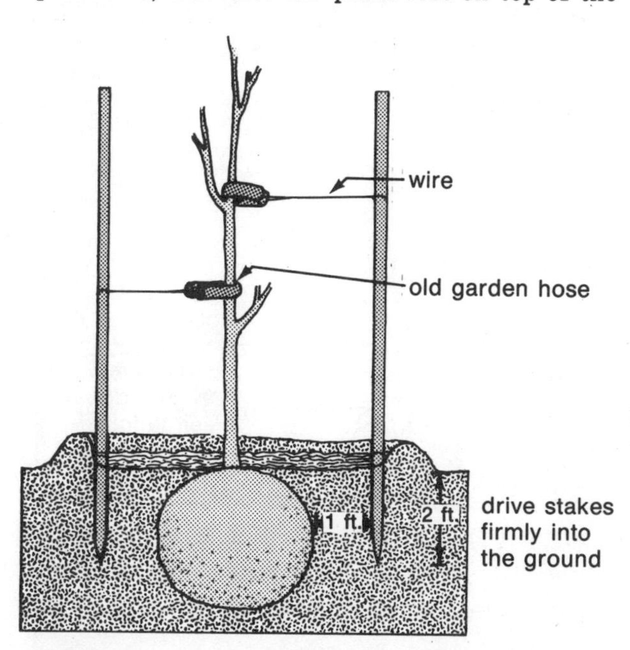

Figure 39. *Supporting transplants with stakes. This method is suitable for light- to medium-branched trees up to 20 feet.*

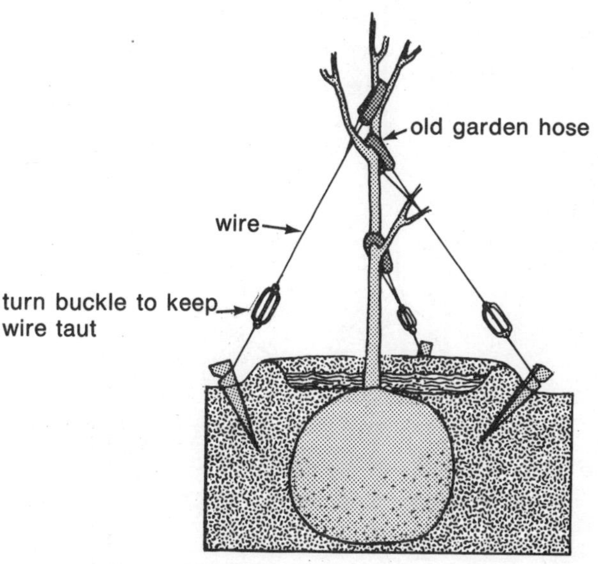

Figure 40. *Guy wire supports should be used to support larger transplants.*

the plant gently upwards. Finally, finish filling to the ground line with the root zone soil mix.

Planting B&B or Container Plants

Place a layer of transition soil in the bottom of the hole as before. Now here is a little secret to really ensure that container plants "take off." Often, these plants are root bound. To relieve this condition and stimulate new root growth, take a sharp knife and score the root ball in several places. With a knife, starting at the top of the ball, make shallow (¼-inch) cuts downward into the root ball about every 4 to 6 inches.

Use the planting board to set the plant at the proper depth. Fill in ⅔ with the root zone soil and tamp gently, then continue as previously described.

Make a Watering Basin

Build a ridge of soil about 6 inches high on the surface of the ground in a circle around the drip line of the tree to form a watering basin.

Support Stake

With all trees and shrubs, it's a good idea to stake them for support against winds (Figures 38-40). A redwood stake 6 to 8 feet long may be driven in about 4 inches or so from the trunk just after the transition zone soil has been placed and before the root zone soil is added. Be sure it is driven in plumb (straight up and down).

Index

Notes

Notes

Notes